Water Quality

PRINCIPLES AND PRACTICES OF WATER SUPPLY OPERATIONS SERIES

Water Sources, Third Edition

Water Treatment,

Water Transmission and Distribution,

Water Quality

Basic Science Concepts and Applications

Water Quality

Third Edition

American Water Works Association

Science and Technology

AWWA unites the entire water community by developing and distributing authoritative scientific and technological knowledge. Through its members, AWWA develops industry standards for products and processes that advance public health and safety. AWWA also provides quality improvement programs for water and wastewater utilities.

Project Manager/Editor: Melissa Christensen
Production Editor/Cover Design: Carol Stearns

Disclaimer

Protection Agency. In no event will AWWA be liable for direct, indirect, special, incidental, or consequential damages arising out of the use of information presented in this book. In particular, AWWA will not be responsible for any costs, including, but not limited to, those incurred as a result of lost revenue. In no event shall AWWA's liability exceed the amount paid for the purchase of this book.

Library of Congress Cataloging-in-Publication Data
Water quality.--3rd ed.
    p. cm. -- (Principles and practices of water supply operations series)
  Includes bibliographical references and index.
  ISBN 1-58321-232-9
    1. Water quality. 2. Water quality--Measurement. I American Water Works Association. II. Series.

TD370.W392 2003
628.1'61--dc22   2003052369

ISBN 1-58321-232-9

American Water Works Association

Contents

Foreword

Water Quality is part four in a five-part series titled Principles and Practices of Water Supply Operations. It contains information required by treatment system operators on drinking water regulations and water quality sampling and monitoring and describes the laboratory equipment and instrumentation used today to analyze drinking water for microbiological, chemical, and physical contaminants.

The other books in the series are

Water Sources
Water Treatment
Water Transmission and Distribution
Basic Science Concepts and Applications (a reference handbook)

References are made to the other books in the series where appropriate in the text.

The reference handbook is a companion to all four books. It contains basic reviews of mathematics, hydraulics, chemistry, and electricity needed for the problems and computations required in water supply operation. The handbook also uses examples to explain and demonstrate many specific problems.

Acknowledgments

This third edition of *Water Quality* has been revised to include the latest available information on new analytical techniques and current federal drinking water regulations. The material has also been reorganized for better coordination with the other books in the series. The author of the revision was Anthony M. Wachinski, PhD, PE. Special thanks go to Bill Lauer and Elise Harrington of the American Water Works Association for their review of the manuscript.

Introduction

Water treatment plant operators are required to understand federal and state laws and the standards that apply to domestic water treatment systems. They should understand how drinking water regulations are administered and why compliance is essential to providing safe drinking water to the public.

Drinking water regulations set the treatment goals for the water supply industry. Their purpose is to ensure uniform delivery of safe and aesthetically pleasing drinking water to the public.

Drinking water regulations specify monitoring requirements, and water system operators are responsible for two types of monitoring: 1) monitoring required to ensure that the water is safe for human consumption, i.e., the water is potable; and 2) monitoring to measure the efficiency of treatment processes.

Water treatment plant operators are responsible for the proper sampling, i.e., the proper collection and preservation, and in some cases, the basic microbiological and chemical analyses of these samples.

This book contains nine chapters. Federal and state regulations continue to become more stringent and complicated. Chapter 1 provides a brief but thorough discussion of the Safe Drinking Water Act and federal drinking water regulations in effect as of publication of this third edition. Information on the regulations and suggested reading sources are provided for additional information.

Each water system operator should have access to the latest state drinking water regulations. This document will detail the specific requirements that must be met and the methods of water system operation, monitoring, and reporting required by the state primacy agency.

Water quality analysis is an important part of the operation of every public water system. Chapter 2 discusses the basics of proper sampling and monitoring.

Large water systems usually have access to complete onsite laboratories. Maintaining an onsite, dedicated laboratory requires a substantial capital investment in equipment and technicians trained to perform the various analyses. Medium size systems often have small laboratories with the capability to perform less complicated analyses. Small systems generally send samples to a state or commercial laboratory for microbiological and chemical analyses. Chapter 3 describes the equipment and instrumentation used in water analyses.

Water system operators are required to perform basic water analyses and to interpret the test results. Chapters 4, 5, 6, 7, and 8 discuss the techniques commonly used to characterize drinking water.

Chapter 9 provides valuable suggestions for customer complaint investigation.
Additional information on equipment, reagents, and detailed test procedures to conduct each test can be found in either of the following references:

- *Standard Methods for the Examination of Water and Wastewater* (most recent edition).
- *Methods of Chemical Analyses for Water and Wastes*, USEPA, Office of Technology (most recent edition).

Simplified procedures for the more common tests are also provided in the following publications:

- AWWA Manual M12, *Simplified Procedures for Water Examination* (most recent edition).
- Several laboratory equipment manufacturers and suppliers have prepared handbooks that outline the required equipment, reagents, and common test procedures.

SELECTED SUPPLEMENTARY READINGS

Manual M12, Simplified Procedures for Water Examination. 2002. Denver, Colo.: American Water Works Association.

Methods of Chemical Analyses of Water and Wastes. 1984. 600/4-79-020. Cincinnati, Ohio: US Environmental Protection Agency.

Standard Methods for the Examination of Water and Wastewater. 20th ed. 1998. A.D. Eaton, L.S. Clesceri, and A.E. Greenberg, eds. Washington D.C.: American Public Health Association, American Water Works Association, and Water Environment Federation.

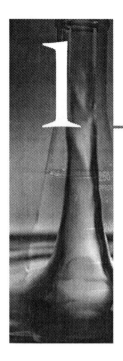

Public Water Supply Regulations

The Safe Drinking Water Act (SDWA), passed by Congress in 1974, started a new era in the field of public water supply. The number of water systems subject to state and federal regulations has vastly increased, and the complexity of the regulations that must now be met far exceeds what could have been imagined just a few years ago. In addition, public water systems are also subject to many additional state and federal environmental and safety regulations.

Under SDWA, USEPA sets legal limits on the levels of certain contaminants in drinking water.

SAFE DRINKING WATER ACT

The principal law governing drinking water safety in the United States is the SDWA. SDWA was passed by Congress and signed into law in 1974. Suspected carcinogens discovered in America's drinking water established a widespread sense of urgency that led to its passage. SDWA directs the US Environmental Protection Agency (USEPA) to promulgate and enforce National Primary Drinking Water Regulations (NPDWRs), which covers 92 contaminants, ensuring safe drinking water for the consumer and protecting public health. These include turbidity, eight microbial or indicator organisms, four radionuclides, 19 inorganic contaminants, and 60 organic contaminants.

Under SDWA, USEPA sets legal limits on the levels of certain contaminants in drinking water. The legal limits reflect both the level that protects human health and the level that water systems can achieve using the best available technology

(BAT). Besides prescribing these legal limits, USEPA rules set water-testing schedules and methods that water systems must follow. The rules also list acceptable techniques for treating contaminated water. SDWA gives individual states the opportunity to set and enforce their own drinking water standards if the standards are at least as strong as USEPA's national standards. Most states and territories directly oversee the water systems within their borders.

The requirements of the SDWA are applicable to all 50 states, the District of Columbia, Indian lands, Puerto Rico, the Virgin Islands, American Samoa, Guam, the Commonwealth of the Northern Mariana Islands, and the Republic of Palau. The intent of the SDWA is for each state to accept primary enforcement responsibility (primacy) for the operation of the state's drinking water program. Indian tribes may also be delegated primacy for administration of public water supplies on tribal lands. As of 1994, all the states and territories except Wyoming and the District of Columbia had accepted primacy. Only a few of the larger Indian tribes have accepted primacy.

SDWA was amended six times between 1974 and 1986 and again in 1996.

The 1986 SDWA amendments set up a timetable under which USEPA was required to develop primary standards for 83 contaminants. Other major provisions required USEPA to

1. Define an approval treatment technique for each regulated contaminant,
2. Specify criteria for filtration of surface water supplies,
3. Specify criteria for disinfecting surface and groundwater supplies,
4. Prohibit the use of lead products in materials used to convey drinking water.

In April 1993, the largest waterborne disease outbreak in the United States occurred in Milwaukee, Wis., when 400,000 people were exposed to the protozoan parasite *Cryptosporidium parvum*. This event attracted national attention to the importance of safe drinking water and influenced the current theme of regulations. The 1996 amendments revised the contaminant list and regulatory process.

On Aug. 6, 1996, new SDWA amendments were signed into Public Law 104-182. These amendments created several new programs and included a total authorization of more than $12 billion in federal funds for drinking water programs.

Public Water Systems

USEPA has further divided public water systems that are covered by SDWA requirements into three categories based on the type of customers served, as follows:

1. *Community public water systems* serve year-round residents and include municipal systems, mobile home parks, apartment buildings with their own water system serving 15 units or more or 25 or more people.

2. *Nontransient, noncommunity public water systems* are entities with their own water supply serving an average of at least 25 persons who do not live at the location but who use the water for more than 6 months per year. These systems include schools and office buildings.

3. *Transient, noncommunity public water systems* are establishments that have their own water system, where an average of at least 25 people per day visit and use the water occasionally or for only short periods of time. Examples include restaurants, hotels, motels, churches, and parks.

A public water system covered under the provisions of the SDWA supplies piped water for human consumption and has at least 15 service connections; or serves 25 or more persons 60 or more days each year.

Examples of systems that do not fall under the provisions of the act are private homes on their own well, housing developments, condominiums, and apartments that each have fewer than 15 connections and serve fewer than 25 residents. Summer camps with a water source that operates fewer than 60 days a year are also included. These systems are usually covered to some degree by state, county, or local health regulations.

Figure 1–1 provides examples of the types of water systems or establishments that are covered under each category. The rationale for dividing systems into these three groups is the chemical exposure of persons using the water. Most chemical contaminants only cause adverse health effects after long-term exposure. Brief exposure of an individual to low levels of a chemical contaminant may not have an effect.

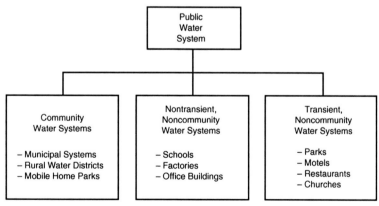

Source: Drinking Water Handbook for Public Officials *(1993).*

FIGURE 1–1 Classification of public water systems

Consider a municipal water system or mobile home park water supply contaminated with a low concentration of a carcinogen, a chemical known to cause cancer. A person who drinks this water every day for a period of years theoretically has an increased chance of getting cancer. A person who works in an office building that has a water supply contaminated by a carcinogen may experience adverse health effects if the person drinks the contaminated water over an extended period of time. A person who visits a hotel and drinks the same contaminated water will only drink a small amount of contaminated water and will have a lower risk of contracting cancer.

Monitoring requirements for community and nontransient, noncommunity water systems apply to all contaminants considered a health threat. Transient, noncommunity systems are only required to monitor for contaminants currently considered to pose a potential health threat from brief exposure, such as nitrite and nitrate, and microbiological contaminants.

Approximately 220,000 public water systems in the United States are regulated under USEPA and SDWA rules. About 60,000 are classed as community systems, and 160,000 fall under one of the two noncommunity systems.

The Water Quality Association categorizes community public water systems according to the number of customers and percentage of population:

- *Very small systems* serve less than 500 people, constitute 64 percent of the community water systems, and serve 2 percent of the community water system population.
- *Small systems* serve 501 to 3,300 customers, constitute 24 percent of the community water systems, and serve 8 percent of the community water system population.
- *Medium systems* serve 3,301 to 10,000 customers, constitute 7 percent of the community water systems, and serve 11 percent of the community water system population.
- *Large systems* serve 10,001 to 100,000 customers, constitute 4.5 percent of the community water systems, and serve 35 percent of the community water system population.
- *Very large systems* serve greater than 100,001 customers, constitute 0.5 percent of the community water systems, and serve 44 percent of the community water system population.

National Primary Drinking Water Regulations

NPDWRs specify maximum Contaminant Levels (MCLs) or treatment techniques (TTs) for contaminants that may have an adverse health effect on humans. The primary regulations are mandatory, and all public water systems must comply with them. If analysis of the water produced by a water system indicates that an

MCL for a contaminant is exceeded, the water system must initiate a treatment regime to reduce the contaminant concentration to below the MCL or take appropriate steps to protect the public's health. Table 1–1 provides the status of USEPA primary drinking water standards at the time this book was prepared, and Table 1–2 provides the required sampling.

Maximum contaminant level goals

The maximum contaminant level goal (MCLG) is the concentration or level of a contaminant in drinking water below which there is no known or expected risk to health. MCLGs allow for a margin of safety and are nonenforceable public health goals. An MCLG is determined using a combination of animal studies and human exposure data. It is the goal the experts would like to see achieved for complete protection of public health.

In some cases the MCLG is economically achievable, and in other instances it is not. For noncarcinogens, the MCLG is a finite number. For known or suspected human carcinogens, the MCLG is zero.

Maximum contaminant levels

The MCL is the highest level of a contaminant allowed in drinking water. MCLs are enforceable standards. SDWA attempts to establish an MCL and an MCLG for each drinking water contaminant. The MCL is set at a level as close as possible to the MCLG but at a concentration that is reasonable and economically achievable with BAT. When it is impossible or impractical to establish an MCL, the USEPA can establish a TT and specify treatment methods that must be used to minimize exposure of the public. Existing MCLs are adjusted from time to time as improved treatment technologies are developed and it becomes economically feasible to move the MCL closer to the MCLG. An MCL may be liberalized if new health effects data indicate so.

Maximum residual disinfectant level goal (MRDLG). MRDLG is the level of a drinking water disinfectant below which there is no known expected risk to health. MRDLGs do not reflect the benefits of a disinfectant used to control microorganisms.

Public Notification

Public water systems that do not comply with the SDWA are required to provide public notification. Systems that violate operating, monitoring, or reporting requirements or briefly exceed an MCL must inform the public of the problem. Even though the problem may have already been corrected, an explanation must be provided in the news media describing the public health significance of the violation.

TABLE 1–1 List of contaminants and their MCLs

Contaminant	MCLG,* mg/L[†]	MCL or TT, mg/L	Potential Health Effects From Ingestion of Water	Sources of Contaminant in Drinking Water
Microorganisms				
Cryptosporidium	zero	TT[‡]	Gastrointestinal illness (e.g., diarrhea, vomiting, cramps)	Human and animal fecal waste
Giardia lamblia	zero	TT	Gastrointestinal illness (e.g., diarrhea, vomiting, cramps)	Human and animal fecal waste
Heterotrophic plate count (HPC)	N/A	TT	HPC has no health effects; it is an analytic method used to measure the variety of bacteria that are common in water. The lower the concentration of bacteria in drinking water, the better maintained the water system is.	HPC measures a range of bacteria that are naturally present in the environment
Legionella	zero	TT	Legionnaire's disease, a type of pneumonia	Found naturally in water; multiplies in heating systems
Total coliforms (including fecal coliform and *E. coli*)	zero	5.0%§	Not a health threat in itself; it is used to indicate whether other potentially harmful bacteria may be present.**	Coliforms are naturally present in the environment; as well as feces; fecal coliforms and *E. coli* only come from human and animal fecal waste.

Table continued next page

6

TABLE 1–1 List of contaminants and their MCLs (continued)

Contaminant	MCLG,* mg/L†	MCL or TT, mg/L	Potential Health Effects From Ingestion of Water	Sources of Contaminant in Drinking Water
Turbidity	N/A	TT	Turbidity is a measure of the cloudiness of water. It is used to indicate water quality and filtration effectiveness (e.g., whether disease-causing organisms are present). Higher turbidity levels are often associated with higher levels of disease-causing microorganisms such as viruses, parasites, and some bacteria. These organisms can cause symptoms such as nausea, cramps, diarrhea, and associated headaches.	Soil runoff
Viruses (enteric)	zero	TT	Gastrointestinal illness (e.g., diarrhea, vomiting, cramps)	Human and animal fecal waste
Disinfection by-products				
Bromate	zero	0.010	Increased risk of cancer	By-product of drinking water disinfection
Chlorite	0.8	1.0	Anemia; infants & young children: Nervous system effects	By-product of drinking water disinfection
Haloacetic acids (HAA5)	N/A††	0.060	Increased risk of cancer	By-product of drinking water disinfection

Table continued next page

TABLE 1-1 List of contaminants and their MCLs (continued)

Contaminant	MCLG,* $mg/L^\dagger$	MCL or TT, mg/L	Potential Health Effects From Ingestion of Water	Sources of Contaminant in Drinking Water
Total trihalomethanes (TTHMs)	none‡‡ / N/A	0.10 / 0.080	Liver, kidney, or central nervous system problems; increased risk of cancer	By-product of drinking water disinfection
Disinfectants				
Chloramines (as Cl_2)	MRDLG =4	MRDL =4.0	Eye/nose irritation; stomach discomfort, anemia	Water additive used to control microbes
Chlorine (as Cl_2)	MRDLG =4	MRDL =4.0	Eye/nose irritation; stomach discomfort	Water additive used to control microbes
Chlorine dioxide (as ClO_2)	MRDLG =0.8	MRDL =0.8	Anemia; infants & young children: Nervous system effects	Water additive used to control microbes
Inorganic Chemicals				
Antimony	0.006	0.006	Increase in blood cholesterol; decrease in blood sugar	Discharge from petroleum refineries; fire retardants; ceramics; electronics; solder
Arsenic	0	0.010 as of 1/23/06	Skin damage or problems with circulatory systems, and may have increased risk of contracting cancer	Erosion of natural deposits; runoff from orchards, runoff from glass and electronics production wastes

Table continued next page

TABLE 1–1 List of contaminants and their MCLs *(continued)*

Contaminant	MCLG,* mg/L†	MCL or TT, mg/L	Potential Health Effects From Ingestion of Water	Sources of Contaminant in Drinking Water
Asbestos (fiber >10 micrometers)	7 million fibers per liter (MFL)	7 MFL	Increased risk of developing benign intestinal polyps	Decay of asbestos cement in water mains; erosion of natural deposits
Barium	2	2	Increase in blood pressure	Discharge of drilling wastes; discharge from metal refineries; erosion of natural deposits
Beryllium	0.004	0.004	Intestinal lesions	Discharge from metal refineries and coal-burning factories; discharge from electrical, aerospace, and defense industries
Cadmium	0.005	0.005	Kidney damage	Corrosion of galvanized pipes; erosion of natural deposits; discharge from metal refineries; runoff from waste batteries and paints
Chromium (total)	0.1	0.1	Allergic dermatitis	Discharge from steel and pulp mills; erosion of natural deposits

Table continued next page

TABLE 1-1 List of contaminants and their MCLs (*continued*)

Contaminant	MCLG,* mg/L†	MCL or TT, mg/L	Potential Health Effects From Ingestion of Water	Sources of Contaminant in Drinking Water
Copper	1.3	TT§§; action level=1.3	Short-term exposure: Gastrointestinal distress Long-term exposure: Liver or kidney damage People with Wilson's disease should consult their personal doctor if the amount of copper in their water exceeds the action level	Corrosion of household plumbing systems; erosion of natural deposits
Cyanide (as free cyanide)	0.2	0.2	Nerve damage or thyroid problems	Discharge from steel/metal factories; discharge from plastics and fertilizer factories
Fluoride	4.0	4.0	Bone disease (pain and tenderness of the bones); children may get mottled teeth	Water additive which promotes strong teeth; erosion of natural deposits; discharge from fertilizer and aluminum factories

Table continued next page

TABLE 1-1 List of contaminants and their MCLs (continued)

Contaminant	MCLG,* mg/L†	MCL or TT, mg/L	Potential Health Effects From Ingestion of Water	Sources of Contaminant in Drinking Water
Lead	zero	TT; action level = 0.015	Infants and children: Delays in physical or mental development; children could show slight deficits in attention span and learning abilities Adults: Kidney problems; high blood pressure	Corrosion of household plumbing systems; erosion of natural deposits
Mercury (inorganic)	0.002	0.002	Kidney damage	Erosion of natural deposits; discharge from refineries and factories; runoff from landfills and croplands
Nitrate (measured as nitrogen)	10	10	Infants below the age of 6 months who drink water containing nitrate in excess of the MCL could become seriously ill and, if untreated, may die. Symptoms include shortness of breath and blue-baby syndrome.	Runoff from fertilizer use; leaching from septic tanks, sewage; erosion of natural deposits

Table continued next page

TABLE 1-1 List of contaminants and their MCLs (*continued*)

Contaminant	MCLG,* $mg/L^\dagger$	MCL or TT, mg/L	Potential Health Effects From Ingestion of Water	Sources of Contaminant in Drinking Water
Nitrite (measured as nitrogen)	1	1	Infants below the age of 6 months who drink water containing nitrite in excess of the MCL could become seriously ill and, if untreated, may die. Symptoms include shortness of breath and blue-baby syndrome.	Runoff from fertilizer use; leaching from septic tanks, sewage; erosion of natural deposits
Selenium	0.05	0.05	Hair or fingernail loss; numbness in fingers or toes; circulatory problems	Discharge from petroleum refineries; erosion of natural deposits; discharge from mines
Thallium	0.0005	0.002	Hair loss; changes in blood; kidney, intestine, or liver problems	Leaching from ore-processing sites; discharge from electronics, glass, and drug factories
Organic Chemicals				
Acrylamide	zero	TT***	Nervous system or blood problems; increased risk of cancer	Added to water during sewage/wastewater treatment
Alachlor	zero	0.002	Eye, liver, kidney, or spleen problems; anemia; increased risk of cancer	Runoff from herbicide used on row crops

Table continued next page

12

TABLE 1–1 List of contaminants and their MCLs (continued)

Contaminant	MCLG,* mg/L†	MCL or TT, mg/L	Potential Health Effects From Ingestion of Water	Sources of Contaminant in Drinking Water
Atrazine	0.003	0.003	Cardiovascular system or reproductive problems	Runoff from herbicide used on row crops
Benzene	zero	0.005	Anemia; decrease in blood platelets; increased risk of cancer	Discharge from factories; leaching from gas storage tanks and landfills
Benzo(a)pyrene (PAHs)	zero	0.0002	Reproductive difficulties; increased risk of cancer	Leaching from linings of water storage tanks and distribution lines
Carbofuran	0.04	0.04	Problems with blood, nervous system, or reproductive system	Leaching of soil fumigant used on rice and alfalfa
Carbon tetrachloride	zero	0.005	Liver problems; increased risk of cancer	Discharge from chemical plants and other industrial activities
Chlordane	zero	0.002	Liver or nervous system problems; increased risk of cancer	Residue of banned termiticide
Chlorobenzene	0.1	0.1	Liver or kidney problems	Discharge from chemical and agricultural chemical factories

Table continued next page

TABLE 1–1 List of contaminants and their MCLs (continued)

Contaminant	MCLG,[*] mg/L[†]	MCL or TT, mg/L	Potential Health Effects From Ingestion of Water	Sources of Contaminant in Drinking Water
2,4-D	0.07	0.07	Kidney, liver, or adrenal gland problems	Runoff from herbicide used on row crops
Dalapon	0.2	0.2	Minor kidney changes	Runoff from herbicide used on rights of way
1,2-Dibromo-3-chloro-propane (DBCP)	zero	0.0002	Reproductive difficulties; increased risk of cancer	Runoff/leaching from soil fumigant used on soybeans, cotton, pineapples, and orchards
o-Dichlorobenzene	0.6	0.6	Liver, kidney, or circulatory system problems	Discharge from industrial chemical factories
p-Dichlorobenzene	0.075	0.075	Anemia; liver, kidney, or spleen damage; changes in blood	Discharge from industrial chemical factories
1,2-Dichloroethane	zero	0.005	Increased risk of cancer	Discharge from industrial chemical factories
1,1-Dichloroethylene	0.007	0.007	Liver problems	Discharge from industrial chemical factories
cis-1,2-Dichloroethylene	0.07	0.07	Liver problems	Discharge from industrial chemical factories

Table continued next page

14

TABLE 1-1 List of contaminants and their MCLs (continued)

Contaminant	MCLG,* mg/L†	MCL or TT, mg/L	Potential Health Effects From Ingestion of Water	Sources of Contaminant in Drinking Water
trans-1,2-Dichloroethylene	0.1	0.1	Liver problems	Discharge from industrial chemical factories
Dichloromethane	zero	0.005	Liver problems; increased risk of cancer	Discharge from drug and chemical factories
1,2-Dichloropropane	zero	0.005	Increased risk of cancer	Discharge from industrial chemical factories
Di(2-ethylhexyl) adipate	0.4	0.4	Weight loss, liver problems, or possible reproductive difficulties	Discharge from chemical factories
Di(2-ethylhexyl) phthalate	zero	0.006	Reproductive difficulties; liver problems; increased risk of cancer	Discharge from rubber and chemical factories
Dinoseb	0.007	0.007	Reproductive difficulties	Runoff from herbicide used on soybeans and vegetables
Dioxin (2,3,7,8-TCDD)	zero	0.00000003	Reproductive difficulties; increased risk of cancer	Emissions from waste incineration and other combustion; discharge from chemical factories
Diquat	0.02	0.02	Cataracts	Runoff from herbicide use
Endothall	0.1	0.1	Stomach and intestinal problems	Runoff from herbicide use

Table continued next page

TABLE 1–1 List of contaminants and their MCLs (continued)

Contaminant	MCLG,* mg/L†	MCL or TT, mg/L	Potential Health Effects From Ingestion of Water	Sources of Contaminant in Drinking Water
Endrin	0.002	0.002	Liver problems	Residue of banned insecticide
Epichlorohydrin	zero	TT	Increased cancer risk, and over a long period of time, stomach problems	Discharge from industrial chemical factories; an impurity of some water treatment chemicals
Ethylbenzene	0.7	0.7	Liver or kidney problems	Discharge from petroleum refineries
Ethylene dibromide	zero	0.00005	Problems with liver, stomach, reproductive system, or kidneys; increased risk of cancer	Discharge from petroleum refineries
Glyphosate	0.7	0.7	Kidney problems; reproductive difficulties	Runoff from herbicide use
Heptachlor	zero	0.0004	Liver damage; increased risk of cancer	Residue of banned termiticide
Heptachlor epoxide	zero	0.0002	Liver damage; increased risk of cancer	Breakdown of heptachlor
Hexachlorobenzene	zero	0.001	Liver or kidney problems; reproductive difficulties; increased risk of cancer	Discharge from metal refineries and agricultural chemical factories

Table continued next page

16

TABLE 1–1 List of contaminants and their MCLs (continued)

Contaminant	MCLG,* mg/L†	MCL or TT, mg/L	Potential Health Effects From Ingestion of Water	Sources of Contaminant in Drinking Water
Hexachlorocyclopentadiene	0.05	0.05	Kidney or stomach problems	Discharge from chemical factories
Lindane	0.0002	0.0002	Liver or kidney problems	Runoff/leaching from insecticide used on cattle, lumber, gardens
Methoxychlor	0.04	0.04	Reproductive difficulties	Runoff/leaching from insecticide used on fruits, vegetables, alfalfa, livestock
Oxamyl (Vydate)	0.2	0.2	Slight nervous system effects	Runoff/leaching from insecticide used on apples, potatoes, and tomatoes
Pentachlorophenol	zero	0.001	Liver or kidney problems; increased cancer risk	Discharge from wood-preserving factories
Picloram	0.5	0.5	Liver problems	Herbicide runoff
Polychlorinated biphenyls (PCBs)	zero	0.0005	Skin changes; thymus gland problems; immune deficiencies; reproductive or nervous system difficulties; increased risk of cancer	Runoff from landfills; discharge of waste chemicals
Simazine	0.004	0.004	Problems with blood	Herbicide runoff

Table continued next page

TABLE 1–1 List of contaminants and their MCLs (*continued*)

Contaminant	MCLG,[*] *mg/L*[†]	MCL or TT, *mg/L*	Potential Health Effects From Ingestion of Water	Sources of Contaminant in Drinking Water
Styrene	0.1	0.1	Liver, kidney, or circulatory system problems	Discharge from rubber and plastics factories; leaching from landfills
Tetrachloroethylene	zero	0.005	Liver problems; increased risk of cancer	Discharge from factories and dry cleaners
Toluene	1	1	Nervous system, kidney, or liver problems	Discharge from petroleum factories
Toxaphene	zero	0.003	Kidney, liver, or thyroid problems; increased risk of cancer	Runoff/leaching from insecticide used on cotton and cattle
2,4,5-TP (Silvex)	0.05	0.05	Liver problems	Residue of banned herbicide
1,2,4-Trichlorobenzene	0.07	0.07	Changes in adrenal glands	Discharge from textile finishing factories
1,1,1-Trichloroethane	0.2	0.2	Liver, nervous system, or circulatory problems	Discharge from metal degreasing sites and other factories
1,1,2-Trichloroethane	0.003	0.005	Liver, kidney, or immune system problems	Discharge from industrial chemical factories

Table continued next page

TABLE 1–1 List of contaminants and their MCLs (continued)

Contaminant	MCLG,* mg/L†	MCL or TT, mg/L	Potential Health Effects From Ingestion of Water	Sources of Contaminant in Drinking Water
Trichloroethylene	zero	0.005	Liver problems; increased risk of cancer	Discharge from metal degreasing sites and other factories
Vinyl chloride	zero	0.002	Increased risk of cancer	Leaching from PVC pipes; discharge from plastics factories
Xylenes (total)	10	10	Nervous system damage	Discharge from petroleum factories; discharge from chemical factories
Radionuclides				
Alpha particles	none / zero	15 picocuries per Liter (pCi/L)	Increased risk of cancer	Erosion of natural deposits of certain minerals that are radioactive and may emit a form of radiation known as alpha radiation
Beta particles and photon emitters	none / zero	4 millirems per year	Increased risk of cancer	Decay of natural and synthetic deposits of certain minerals that are radioactive and may emit forms of radiation known as photons and beta radiation

Table continued next page

19

TABLE 1-1 List of contaminants and their MCLs (continued)

Contaminant	MCLG,[*] mg/L[†]	MCL or TT, mg/L	Potential Health Effects From Ingestion of Water	Sources of Contaminant in Drinking Water
Beta particles and photon emitters	none ___ zero	4 millirems per year	Increased risk of cancer	Decay of natural and synthetic deposits of certain minerals that are radioactive and may emit forms of radiation known as photons and beta radiation
Radium 226 and radium 228 (combined)	none ___ zero	5 pCi/L	Increased risk of cancer	Erosion of natural deposits
Uranium	zero	30 µg/L as of 12/8/03	Increased risk of cancer, kidney toxicity	Erosion of natural deposits

*Definitions:

Maximum contaminant level (MCL)—The highest level of a contaminant that is allowed in drinking water. MCLs are set as close to MCLGs as feasible using the best available treatment technology and taking cost into consideration. MCLs are enforceable standards.
Maximum contaminant level goal (MCLG)—The level of a contaminant in drinking water below which there is no known or expected risk to health. MCLGs allow for a margin of safety and are nonenforceable public health goals.
Maximum residual disinfectant level (MRDL)—The highest level of a disinfectant allowed in drinking water. There is convincing evidence that addition of a disinfectant is necessary for control of microbial contaminants.
Maximum residual disinfectant level goal (MRDLG)—The level of a drinking water disinfectant below which there is no known or expected risk to health. MRDLGs do not reflect the benefits of the use of disinfectants to control microbial contaminants.
Treatment Technique (TT)— A required process intended to reduce the level of a contaminant in drinking water.
†Units are in milligrams per liter (mg/L) unless otherwise noted. Milligrams per liter is equivalent to parts per million.

Table continued next page

TABLE 1–1 List of Contaminants and Their MCLs (continued)

Contaminant	MCLG (mg/L)	MCL or TT (mg/L)	Potential Health Effects From Ingestion of Water	Sources of Contaminant in Drinking Water

‡USEPA's Surface Water Treatment Rules (SWTRs) require systems using surface water or groundwater under the direct influence of surface water to (1) disinfect their water, and (2) filter their water or meet criteria for avoiding filtration so that the following contaminants are controlled at the following levels:

- *Cryptosporidium* (as of 1/1/02 for systems serving >10,000 and 1/14/05 for systems serving <10,000) 99% removal.
- *Giardia lamblia*: 99.9% removal/inactivation.
- Viruses: 99.99% removal/inactivation.
- *Legionella*: No limit, but USEPA believes that if *Giardia* and viruses are removed/inactivated, *Legionella* will also be controlled.
- Turbidity: At no time can turbidity (cloudiness of water) go above 5 nephelometric turbidity units (ntu); systems that filter must ensure that the turbidity go no higher than 1 ntu (0.5 ntu for conventional or direct filtration) in at least 95% of the daily samples in any month. As of Jan. 1, 2002, turbidity may never exceed 1 ntu and must not exceed 0.3 ntu 95% of daily samples in any month.
- HPC: No more than 500 bacterial colonies per milliliter.
- Long-Term 1 Enhanced Surface Water Treatment Rule (effective date: Jan. 14, 2005): Surface water systems or (GWUDI) systems serving fewer than 10,000 people must comply with the applicable Long-Term 1 Enhanced Surface Water Treatment Rule provisions (e.g., turbidity standards, individual filter monitoring, *Cryptosporidium* removal requirements, updated watershed control requirements for unfiltered systems).
- Filter Backwash Recycling: The Filter Backwash Recycling Rule requires systems that recycle to return specific recycle flows through all processes of the system's existing conventional or direct filtration system or at an alternate location approved by the state.

§More than 5.0% samples total coliform-positive in a month. (For water systems that collect fewer than 40 routine samples per month, no more than one sample can be total coliform-positive per month.) Every sample that has total coliform must be analyzed for either fecal coliforms or *E. coli*; if two consecutive total coliform positive samples and one is also positive for *E. coli* fecal coliforms, system has an acute MCL violation.

Table continued next page

TABLE 1-1 List of Contaminants and Their MCLs (continued)

Contaminant	MCLG (mg/L)	MCL or TT (mg/L)	Potential Health Effects From Ingestion of Water	Sources of Contaminant in Drinking Water

**Fecal coliform and *E. coli* are bacteria whose presence indicates that the water may be contaminated with human or animal wastes. Disease-causing microbes (pathogens) in these wastes can cause diarrhea, cramps, nausea, headaches, or other symptoms. These pathogens may pose a special health risk for infants, young children, and people with severely compromised immune systems.

††Although there is no collective MCLG for this contaminant group, there are individual MCLGs for some of the individual contaminants:

- Trihalomethanes: bromodichloromethane (zero); bromoform (zero); dibromochloromethane (0.06 mg/L). Chloroform is regulated with this group but has no MCLG.
- Haloacetic acids: dichloroacetic acid (zero); trichloroacetic acid (0.3 mg/L). Monochloroacetic acid, bromoacetic acid, and dibromoacetic acid are regulated with this group but have no MCLGs.

‡‡MCLGs were not established before the 1986 amendments to the SDWA. Therefore, there is no MCLG for this contaminant.

§§Lead and copper are regulated by a treatment technique that requires systems to control the corrosiveness of their water. If more than 10% of tap water samples exceed the action level, water systems must take additional steps. For copper, the action level is 1.3 mg/L, and lead is 0.015 mg/L.

***Each water system must certify, in writing, to the state (using third-party or manufacturer's certification) that when acrylamide and epichlorohydrin are used in drinking water systems, the combination (or product) of dose and monomer level does not exceed the levels specified, as follows:

- Acrylamide = 0.05% dosed at 1 mg/L (or equivalent)
- Epichlorohydrin = 0.01% dosed at 20 mg/L (or equivalent)

TABLE 1-2 Required sampling

Required Tests	Sampling Location	Frequency: Community and Nontransient, Noncommunity Systems	Frequency: Transient, Noncommunity Systems
Inorganics	Entry points to distribution system.	Systems using surface water: every year. Systems using groundwater only: every 3 years.	Nitrate: yearly. Nitrite: at state option.
Organics: except trihalomethanes	Entry points to distribution system.	Systems using surface water: every 3 years. Systems using groundwater only: state option.	State option.
Organics: trihalomethanes	25% at extremes of distribution system; 75% at locations representative of population distribution.	Systems serving populations of 10,000 or more: four samples per quarter per plant.*	State option.
Turbidity	At point(s) where water enters distribution system.	Systems using surface water: daily. Systems using groundwater only: state option.	Systems using surface water or surface water and groundwater only: daily. Systems using groundwater only: state option.
Coliform bacteria	At consumer's faucet.	Depends on number of people served by water system.	Systems using surface water and/ or groundwater: one per quarter (for each quarter water is served to public).
Radiochemicals: natural	At consumer's faucet.	Systems using surface water: every 4 years. Systems using groundwater only: state option.	State option.
Radiochemicals: synthetic	At consumer's faucet (at state option).	Systems using surface water serving populations greater than 100,000: every 4 years. All other systems: state option.	System using surface water and/ or groundwater: state option.

*Systems using multiple wells drawing raw water from a single aquifer may, with state approval, be considered one treatment plant for determining the required number of samples.

Some violations are more serious than others and two tiers of public notification have been established (Tables 1–3 and 1–4). Tier 1 violations are more serious than tier 2 violations and have more extensive notification requirements.

USEPA provides language mandatory for use with each type of public notification to fully inform the public of the significance of the violation. Violations of *state* reporting requirements may not require public notification, depending on the infraction and state policy.

Formal Enforcement

In instances of serious or prolonged noncompliance with federal requirements, SDWA has provided USEPA and the states with authority to assess stiff monetary fines.

TABLE 1–3 Summary of notification requirements

Category of Violation	Mandatory Health Effects Information Required (all public water supplies)	Notice to New Billing Units (community water supplies only)
Tier 1		
Maximum contaminant level	Yes	Yes
Treatment technique	Yes	Yes
Variance of exemption schedule violation	Yes	Yes
Tier 2		
Monitoring[*]	No	No
Testing procedures	No	No
Variance of exemption issued	Yes	No

Source: Adapted from Public Notification Handbook for Public Officials *(USEPA 1989).*

*Continuous report required if posting is used; quarterly report required if hand delivery is used.

TABLE 1–4 Types of notification and time frames for notification of community and noncommunity water supplies

Type of Public Water Supply	Types of Violation and Notification Method	Time Frame Within Which Notice Must Be Given (X indicates time frame for initial notice)					Frequency of Repeat Notices Until Violation Is Resolved
		72 hours	7 days	14 days	45 days	3 months	
Community	Tier 1: Acute violations						
	• TV and radio	X					No repeat
	• Newspaper*			X			No repeat
	• Mail or hand delivery†				X		Quarterly repeat
Community	Tier 1: Nonacute violations						
	• Newspaper*			X			No repeat
	• Mail or hand delivery				X		Quarterly repeat
Community	Tier 2: All violations						
	• Newspaper*					X	Quarterly repeat by mail or hand delivery
Noncommunity‡	Tier 1: Acute violations						
	• Option 1: Notice as for community water systems (above)						

Table continued next page

TABLE 1–4 Types of notification and time frames for notification of community and noncommunity water supplies (continued)

Type of Public Water Supply	Types of Violation and Notification Method	Time Frame Within Which Notice Must Be Given (X indicates time frame for initial notice)					Frequency of Repeat Notices Until Violation Is Resolved
		72 hours	7 days	14 days	45 days	3 months	
	• Option: 2: Posting or hand delivery	X					Continuous or quarterly repeat
Noncommunity	Tier 1: Nonacute violations						
	• Option 1: Notice as for community water systems (above)						
	• Option 2: Posting or hand delivery			X			Continuous or quarterly repeat
Noncommunity	Tier 2: All violations						
	• Option 1: Notice as for community water system (above)						
	• Option 2: Posting or hand delivery					X	Continuous or quarterly repeat§

Source: Adapted from *Public Notification Handbook for Public Officials (USEPA 1989).*

*If no newspaper of general circulation is available, posting or hand delivery is required.
†May be waived.
‡Includes both transient, noncommunity public water systems and nontransient, noncommunity public water systems.
§Less frequent notice (but no less than annual) to be required.

Monitoring and Reporting Requirements

To ensure that drinking water meets federal and state requirements, all water systems are required to regularly sample and test the water supplied to consumers. The regulations specify minimum sampling frequencies, sampling locations, testing procedures, requirements for record keeping, and routine reporting to the state. The regulations also cover special reporting procedures to be followed if a contaminant exceeds an MCL.

Monitoring

The federal regulations specify minimum monitoring frequencies, which in many cases are a function of the type of water source being used, the type of treatment used, and the size of the water system. All systems must provide periodic testing for microbiological contamination and analysis for some chemical contaminants.

With the continual addition of new requirements for further testing of water quality, USEPA has instituted a reorganization of monitoring requirements called the *standardized monitoring framework*.

Reporting and record keeping

The results of all water analyses must be provided periodically to the state. Failure to have the proper analyses performed or to report the results to the state primacy agency will usually result in the water system having to provide public notification. Specific information, shown in Table 1–5, must be included on every laboratory report.

There are also specific requirements for the records that must be kept by water systems on their operation and monitoring and for the length of time the records must be retained. These requirements are summarized in Table 1–6. Although state requirements for monitoring, reporting, and record retention must be as stringent as federal requirements, they often vary and may include specific procedures that must be used.

Variances and Exemptions

Each drinking water regulation includes provisions for variances and exemptions. States are authorized to grant one or more variances to a water system that cannot comply with an MCL because of characteristics of the water source(s). A variance may only be granted to systems that have installed full-scale BAT for treatment of the MCL being violated. Granting of a variance must not result in an unreasonable risk to the public health, and the state must prescribe a schedule of compliance.

TABLE 1-5 Laboratory report summary requirements

Type of Information	Summary Requirement
Sampling information	Date, place, and time of sampling
	Name of sample collector
	Identification of sample
	• Routine or check sample
	• Raw or treated water
Analysis information	Date of analysis
	Laboratory conducting analysis
	Name of person responsible
	Analytical method used
	Analysis results

TABLE 1-6 Record-keeping requirements

Type of Record	Time Period
Bacteriological and turbidity analyses	5 years
Chemical analyses	10 years
Actions taken to correct violations	3 years
Sanitary survey reports	10 years
Exemptions	5 years following expiration

States may exempt a water system from an MCL or treatment technique requirement if it finds that all three of the following conditions exist:

1. The system is unable to comply with the requirement because of compelling factors (which may include economic factors).
2. The exemptions will not result in an unreasonable risk to public health.

3. The system was in operation as of Jan. 1, 1989, or, if it was not, no reasonable alternative source of drinking water is available to the new system.

National Secondary Drinking Water Regulations

A National Secondary Drinking Water Regulation is a nonenforceable guideline regarding contaminants that may cause cosmetic effects such as taste, odor, and color. Some states choose to adopt them as enforceable standards. Table 1–7 lists the secondary MCLs. Table 1–8 lists the adverse effects of secondary contaminants.

TABLE 1–7 National Secondary Drinking Water Regulations

Contaminant	Secondary Standard
Aluminum	0.05 to 0.2 mg/L
Chloride	250 mg/L
Color	15 (color units)
Copper	1.0 mg/L
Corrosivity	noncorrosive
Fluoride	2.0 mg/L
Foaming agents	0.5 mg/L
Iron	0.3 mg/L
Manganese	0.05 mg/L
Odor	3 threshold odor number
pH	6.5–8.5
Silver	0.10 mg/L
Sulfate	250 mg/L
Total dissolved solids	500 mg/L
Zinc	5 mg/L

NOTE: For more information, read *Secondary Drinking Water Regulations: Guidance for Nuisance Chemicals.*

TABLE 1–8 Adverse effects of secondary contaminants

Contaminant	Adverse Effect
Chloride	Causes taste. Adds to total dissolved solids and scale. Indicates contamination.
Color	Indicates dissolved organics may be present, which may lead to trihalomethane formation. Unappealing appearance.
Copper	Undesirable metallic taste.
Corrosivity	Corrosion products unappealing to consumers. Causes tastes and odors. Corrosion products can affect health. Corrosion causes costly deterioration of water system.
Fluoride	Dental fluorosis (mottling or discoloration of teeth).
Foaming agents	Unappealing appearance. Indicates possible contamination.
Hydrogen sulfide	Offensive odor. Causes black stains on contact with iron. Can accumulate to deadly concentration in poorly ventilated areas. Flammable and explosive.
Iron	Discolors laundry brown. Changes taste of water, tea, coffee, and other beverages.
Manganese	Discolors laundry. Changes taste of water, tea, coffee, and other beverages.
Odor	Unappealing to drink. May indicate contamination.
pH	Below 6.5, water is corrosive. Above 8.5, water will form scale, taste bitter.
Sulfate	Has a laxative effect.
Total dissolved solids	Associated with taste, scale, corrosion, and hardness.
Zinc	Undesirable taste. Milky appearance.

CURRENT AND FUTURE RULES AFFECTING DRINKING WATER SYSTEMS

Existing rules intended to control microbial risks include the following:

Total Coliform Rule (TCR)
Unregulated Contaminant Monitoring Rule (UCMR)
Surface Water Treatment Rule (SWTR)
Filter Backwash Recycling Rule (FBRR)

Existing rules intended to control chemical risks include the following:

National Interim Primary Drinking Water Regulations (NIPDWRs)
Fluoride Rule (FR)
Volatile Organic Chemicals (Phase I) (VOCs)
Lead and Copper Rule (LCR)
Synthetic Organic Chemicals and Inorganic Chemicals (Phase II) (SOCs & IOCs)
Stage I Disinfectants/Disinfection By-Products Rule (D/DBPR)
Radionuclides Rule
Consumer Confidence Report (CCR) Rule
Public Notification (PN) Rule
Unregulated Contaminant Monitoring Rule (UCMR)
Future rules intended to control microbial risks include the following:
Long-Term 1 Enhanced Surface Water Treatment Rule (LT1ESWTR)
Long-Term 2 ESWTR
Ground Water Rule (GWR)
Future rules intended to control chemical risks include the following:
Arsenic
Radon
Stage 2 Disinfection By-Products (DBPs)
NPDWR Revisions
Drinking Water Contaminant Candidate List (DWCCL)

Total Coliform Rule

The Total Coliform Rule (TCR) (published 29 June 1989/effective 31 December 1990) set both health goals (MCLGs) and legal limits (MCLs) for total coliform levels in drinking water. The rule also details the type and frequency of testing that water systems must perform.

The coliforms are a broad class of bacteria, which live in the digestive tracts of humans and many animals. The presence of coliform bacteria in tap water suggests that the treatment system is not working properly or that there is a problem in the pipes. Among the health problems that contamination can cause are diarrhea, cramps, nausea, and vomiting. Together these symptoms comprise a general category known as gastroenteritis. Gastroenteritis is not usually serious for a healthy person, however, it can lead to more serious problems for people with weakened immune systems, such as the very young, elderly, or immunocompromised.

In the rule, USEPA set the health goal for total coliforms at zero. Because there have been waterborne disease outbreaks in which researchers have found very low levels of coliforms, any level indicates some health risk.

USEPA also set a legal limit on total coliforms. Systems must not find coliforms in more than 5 percent of the samples they take each month to meet USEPA's standards. If more than 5 percent of the samples contain coliforms, water system operators must report this violation to the state and the public.

When a system finds coliforms in drinking water, it may indicate that the system's treatment system is not performing properly. To avoid or eliminate microbial contamination, systems may need to take a number of actions, including repairing the disinfection/filtration equipment, flushing or upgrading the distribution system, and enacting source water protection programs to prevent contamination.

If a sample tests positive for coliforms, the system must collect a set of repeat samples within 24 hours. When a routine or repeat sample tests positive for total coliforms, it must also be analyzed for fecal coliforms and *Escherichia coli (E. coli)*, which are coliforms directly associated with fresh feces. A positive result to this last test signifies an acute MCL violation, which necessitates rapid state and public notification because it represents a direct health risk.

The number of coliform samples a system must take depends on the number of customers that it serves. Systems that serve fewer than 1,000 people may test once a month or less frequently, while systems with 50,000 customers must test 60 times per month, and those with 2.5 million customers must test at least 420 times per month. These are minimum schedules, and many systems test more frequently.

Unregulated Contaminant Monitoring Rule

The 1996 amendments to SDWA requires USEPA to establish criteria for a monitoring program for unregulated contaminants and to publish a list of contaminants to be monitored. USEPA has revised the Unregulated Contaminant Monitoring Rule (UCMR). The data generated by the new UCMR will be used to evaluate and prioritize contaminants on the Drinking Water Contaminant Candidate List (DWCCL), a list of contaminants USEPA is considering for possible new drinking water standards. These data will help to ensure that future decisions on drinking water standards are based on sound science.

The new rule includes

- a new list of contaminants for which public water systems must monitor;
- analytical methods for some of these contaminants;
- requirements for all large public water systems (PWSs), and a representative sample of small PWSs, to monitor for those contaminants on the list for which methods have been promulgated;
- requirements to submit the monitoring data to USEPA and the states for inclusion in the National Drinking Water Contaminant Occurrence Database;
- requirements to notify consumers of the results of monitoring.

USEPA revised the UCMR on Sept. 17, 1999.

USEPA uses the data generated by the new UCMR to evaluate and prioritize contaminants on the DWCCL, a list of contaminants USEPA is considering for possible new drinking water standards. These data help to ensure that USEPA has the high-quality scientific data it needs to make decisions about future drinking water standards.

Surface Water Treatment Rule

The Surface Water Treatment Rule (published 29 June 1989/effective 31 December 1990) seeks to prevent waterborne diseases caused by viruses, *Legionella*, and *Giardia lamblia*. These disease-causing microbes are present at varying concentrations in most surface waters. The rule requires that water systems filter and disinfect water from surface water sources to reduce the occurrence of unsafe levels of these microbes.

The rule applies to the operation of every public water system that uses surface water as a source. It also imposes new requirements on water systems that use groundwater that might become contaminated by surface water; these systems are termed groundwater under the direct influence of surface water, usually abbreviated GWUDI.

As the title suggests, this rule governs water supplies whose source of drinking water is surface water, which it defines as "all water open to the atmosphere and subject to surface runoff." This water, which most of the country's large water systems use, is in rivers, lakes, and reservoirs. Surface water is particularly susceptible to microbial contamination from sewage treatment plant discharges and runoff from stormwater and snowmelt. These sources often contain high levels of fecal microbes that originated in livestock wastes or septic systems.

The purpose of the regulation was to protect the public from waterborne disease. The organisms that cause the waterborne diseases most frequently diagnosed in the United States are *G. lamblia*, *Cryptosporidium*, *Legionella*, viruses, and some types of bacteria. No simple, inexpensive tests are available for detecting the presence of *Cryptosporidium*, *Giardia*, and *Legionella*. Current methods for determining coliform are only general indicators of fecal contamination, not really true indicators of the presence of the other types of organisms.

Because of this inability to test routinely for the presence of specific microorganisms, USEPA has required the use of a "treatment technique" by all surface water systems. The treatment used must ensure that the finished water will meet the water quality goals without the need for specific testing. Studies indicate that *Cryptosporidium* oocysts, *Giardia* cysts, and viruses are among the most resistant waterborne pathogens. Surface water systems must therefore utilize filtration and disinfection processes that will either remove or inactivate virtually all of these microorganisms. Treatment must assure the removal or inactivation of 99.9 percent (3 logs) of *Giardia* cysts and 99.99 percent (4 logs) of viruses.

All systems must filter and disinfect their water to provide a minimum of 99.9 percent combined removal and inactivation of *Giardia* and 99.99 percent of viruses. The adequacy of the filtration process is established by measuring turbidity (a measure of the concentration of particles) in the treated water and determining if it meets USEPA's performance standard. Some public water supplies that have pristine sources may be granted a waiver from the filtration requirement. These supplies must provide the same level of treatment as those that filter; however, their treatment is provided through disinfection alone. The great majority of water supplies in the United States that use a surface water source filter their water.

To assure adequate microbial protection in the distribution system, water systems are also required to provide continuous disinfection of the drinking water entering the distribution system and to maintain a detectable disinfectant level within the distribution system. The distribution system is a series of pipes that delivers treated water from the water treatment plant to the consumer's tap.

Ingestion of *Cryptosporidium*, *Giardia* (a protozoa), and viruses can cause problems in the human digestive system, generally in the form of diarrhea, cramps, and nausea. *Legionella* bacteria in water are only a health risk if the bacteria are aerosolized (e.g., in an air-conditioning system or a shower) and then inhaled. Inhalation can result in a type of pneumonia known as Legionnaires' disease.

The rule sets nonenforceable health goals, or MCLGs, for *Legionella*, *Giardia*, and viruses at zero because any amount of exposure to these contaminants represents some health risk.

C × T values

The SWTR treatment technique goals may be partially met by using disinfection treatment. The effectiveness of a disinfectant in inactivating *Giardia* cysts and viruses depends on

- type of disinfectant used,
- residual concentration of the disinfectant,
- period of time the water is in contact with the disinfectant,
- water temperature,
- chlorine used, and
- pH of the water.

USEPA has determined that a combination of the residual concentration, *C*, of a disinfectant (in milligrams per liter) multiplied by the contact time, *T* (in minutes), can be used as a measure of the disinfectant's effectiveness in killing or inactivating microorganisms. In other words, a water system can use a relatively small application of disinfectant and keep it in contact with the water for a long time or use a large disinfectant dose in contact with the water for a short time and obtain approximately the same results.

The requirement for all surface water systems without filtration treatment is that they must compute the $C \times T$ value for their treatment process daily, and it must always be above the minimum value specified by USEPA. The allowable levels vary by both the type(s) of disinfectant used and the water temperature. Systems using filtration treatment must calculate and meet the $C \times T$ values specified by the state primacy agency.

Filtration treatment

Most surface water systems and systems that are designated by the state as GWUDI must provide both disinfection and filtration treatment to meet the treatment technique requirements. The current technologies specified by the SWTR are conventional treatment, direct filtration, slow sand filtration, diatomaceous earth filtration, reverse osmosis, and alternate technologies, which include other membrane technologies and ultraviolet (UV). These processes are covered in detail in another book in this series, *Water Treatment.*

Some surface water systems that are using especially clean and protected water sources can avoid the requirement to provide filtration and may use disinfection treatment only. However, these systems must meet many additional requirements for providing source water protection and monitoring the operation of their system and the water quality.

Other SWTR requirements

Some of the other principal requirements of the SWTR are as follows:

- For most systems, the turbidity of water entering the distribution system must be equal to or less than 0.5 nephelometric turbidity units (ntu) in at least 95 percent of the measurements taken each month. At no time may the turbidity exceed 5 ntu.

- The disinfection residual of water entering the distribution system must generally be monitored continuously for systems serving a population of more than 3,300. The residual cannot be less than 0.2 mg/L for more than 4 hours during periods when water is being served to the public. Any time the residual falls below this level, the system must notify the state.

- The disinfectant residual must be measured at the same points on the distribution system that are used for coliform sampling. Disinfectant residuals must not be undetectable in more than 5 percent of the samples each month for any 2 consecutive months that water is served to the public.

- Systems must submit special reports to the state detailing the monitoring required by the SWTR.

Filter Backwash Recycling Rule

The Filter Backwash Recycling Rule (FBRR) is intended to reduce the opportunity for recycle practices to adversely effect the performance of drinking water treatment plants and to help prevent microbes, such as *Cryptosporidium*, from passing through treatment systems and into finished drinking water. Customers may become ill if they drink such contaminated water.

Spent filter backwash water, thickener supernatant, and liquids from dewatering processes can contain microbial contaminants, often in very high concentrations. Recycling these streams can reintroduce microbes and other contaminants to the treatment system. Additionally, large volumes of recycle streams may upset treatment processes, allowing contaminants to pass through the system. To minimize these risks, the FBRR requires that recycle streams pass through all the processes of a system's existing conventional or direct filtration system (as defined in 40 CFR 141.2) that USEPA has recognized as capable of achieving 2-log (99 percent) *Cryptosporidium* removal. The FBRR also allows recycle streams to be reintroduced at an alternate location, if the location is state approved.

Lead and Copper Rule

The Lead and Copper Rule (published 7 June 1991/effective 7 December 1992) is substantially different from the rest of this cluster of rules. The other rules require water systems to treat water so that when it leaves their facilities, it is clean and safe to drink. This rule regulates two contaminants that nearly always taint drinking water after it leaves the treatment plant.

Lead and copper are both naturally occurring metals. Both have been used to make household plumbing fixtures and pipes for many years, though Congress banned the use of lead solder, pipes, and fittings in 1986. The two contaminants enter drinking water when water reacts with the metals in the pipes. This is likely to happen when water sits in a pipe for more than a few hours.

Lead and copper have different health effects. Lead is particularly dangerous to fetuses and young children because it can slow their neurological and physical development. Anemia may be one sign of a child's exposure to high lead levels. Lead may also affect the kidneys, brain, nervous system, and red blood cells. It is considered a possible cause of cancer. Copper is a health concern for several reasons. At very low levels, it is necessary to the body, but, in the short term, consumption of drinking water containing copper well above the action level could cause nausea, vomiting, and diarrhea. It can also lead to serious health problems in people with Wilson's disease. Exposure to drinking water containing copper above the action level over many years could increase the risk of liver and kidney damage. To prevent these effects, USEPA set health goals (MCLGs) and action levels for lead and copper.

USEPA required water systems to evaluate not only the pipes in their distribution systems but also the age and types of housing that they serve. Based on this information, the systems must collect water samples at points throughout the distribution system that are vulnerable to lead contamination, including regularly used bathroom or kitchen taps.

When the level of lead or copper reaches the action level in 10 percent of the tap water samples, the water system must begin certain water treatment steps. An action level is different from an MCL in that while an MCL is a legal limit on a contaminant, an action level is a trigger for additional prevention or removal steps.

The rule requires water systems to apply certain treatment techniques for high lead or copper levels. At a minimum, systems must maintain optimal corrosion control. Corrosion control does not reduce the contaminant level but helps prevent the water from being contaminated in the first place. By increasing the water's pH or hardness, water systems can make their water less corrosive, and therefore less likely to corrode the pipes and absorb the lead or copper. Consumers can further reduce the potential for elevated lead levels at the tap by ensuring that all plumbing and fixtures meet local plumbing codes.

When a water system exceeds either action level, it must also assess its source water. In most cases, there will be little or none of either contaminant in the source water and no treatment will be necessary. When there are high levels in the source water, treatment of that water, in conjunction with corrosion control, further lessens the chance that consumers will have elevated levels of lead and copper at the tap.

The rule also requires systems that exceed the lead action level to educate the affected public about reducing its lead intake.

Public Notification Rule

Owners and operators of public water systems are required to notify persons served when they fail to comply with requirements of the NPDWR and pose a risk to public health. Final regulations to revise the general public notification regulations for public water systems were published on May 4, 2000. These regulations were required by the 1996 SDWA Amendments.

DRINKING WATER PROGRAM REQUIREMENTS

Reporting and Record Keeping
The results of all water analyses must be provided periodically to the state. Failure to have the proper analyses performed or to report the results to the state primacy agency will usually result in the water system having to provide public notification. Specific information, shown in Table 1–5, must be included on every laboratory report.

There are also specific requirements for the records that must be kept by water systems on their operation and monitoring and for the length of time the records must be retained. These requirements are summarized in Table 1–6. Although state requirements for monitoring, reporting, and record retention must be as stringent as the federal requirements, they often vary and may include specific procedures that must be used.

SPECIAL REGULATION REQUIREMENTS
In addition to the initial programs required by the SDWA, USEPA has drafted several specific rules to address various types of water contaminant problems. Some of these rules have been promulgated, and others are still under development. Several of the more important rules are described in the following sections.

Optimal Corrosion Control
Large utilities must conduct corrosion control studies unless they can demonstrate that their corrosion control is already optimal. Utilities have attained optimal corrosion control if the difference between the source water lead concentration and the ninetieth percentile from the tap samples is less than 0.005 mg/L.

Small and medium-size utilities are deemed to have optimal corrosion control if they meet the lead and copper action levels for two consecutive sampling periods. How to conduct corrosion control studies is outlined in references listed in the Selected Supplementary Readings at the end of this chapter.

Water Quality Parameters
All large water systems, as well as all medium-size and small water systems that exceed the action levels, are required to monitor for additional contaminants—a practice that will help determine if the systems are maintaining optimal corrosion control. These parameters are analyzed as follows:
- Conductivity may be measured in the field or the sample returned to the laboratory for measurement.

- The pH must be measured in the field, and only the probe method is approved by USEPA.

- Temperature must be measured in the field along with pH and may be measured with a handheld thermometer or with a combined temperature–pH electrode.

- Calcium must be measured in the laboratory. Because the sample for calcium must be acidified for analysis, a separate sample for calcium must be collected.

- Alkalinity must be measured in the laboratory.

- If a phosphate-based corrosion inhibitor is used, an orthophosphate analysis must be conducted.

- If a silica-based corrosion inhibitor is used, a silica analysis must be conducted.

Samples for the water quality parameters may be collected at the usual bacterial sample points in the distribution system. Either glass or plastic containers may be used unless silica is being measured, in which case plastic is required. Samples should be collected from fully flushed sample taps, and the representativeness of the sample site for producing the desired data must be considered.

Results of these analyses are reviewed by the primacy agency, which will then establish the ranges of the parameters within which the utility may operate.

Arsenic Rule

At the time of this writing, USEPA is finalizing a regulation to reduce the public health risks from arsenic in drinking water. It is revising the current drinking water standard for arsenic from 50 parts per billion (ppb) to 10 ppb. This revision will provide additional protection for 13 million Americans against cancer and other health problems, including cardiovascular disease and diabetes, as well as neurological effects. USEPA is reviewing the new arsenic in drinking water standard. It will work with the National Academy of Sciences and the National Drinking Water Advisory Council to reassess the scientific and cost issues associated with the rule.

Stage 1 Disinfectants/Disinfection By-Products Rule (Stage 1 D/DBPR)

Disinfection by-products are formed when organic materials, naturally present in source waters, combine with a disinfectant. Common organics present in many surface water sources are humic acids. Disinfectants commonly used in drinking water treatment include chlorine, chloramines, ozone, and chlorine dioxide. The amount and type of DBPs formed depend on many factors, including the amount and type of organic precursors initially present, pH, time of exposure to disinfectant, temperature, type of disinfectant, plus other factors.

The Disinfectants/Disinfection By-Products (D/DBP) Rule applies to both surface and groundwater systems and has far-reaching effects for US water utilities. Unlike the existing total trihalomethane (TTHM) regulation that only applies to systems serving over 10,000 people, the D/DBP Rule applies to all systems, regardless of the size of the population served. Stage 1 of the rule lowers the allowable TTHM levels from 100 µg/L to 80 µg/L. In addition, it will regulate haloacetic acids (HAA5) at 60 µg/L chlorite at 1.0 mg/L, and bromate at 0.010 mg/L. The rule sets maximum residual disinfectant levels for chlorine (4 mg/L as Cl_2), chloramines (4.0 mg/L as Cl_2), and chlorine dioxide (0.8 mg/L as ClO_2). The rule also includes a treatment requirement for "enhanced coagulation" for surface water sources where conventional treatment is applied. The time line for compliance ranges from 18 months for surface water systems serving 10,000 or more people; up to 60 months for groundwater systems serving less than 10,000 persons.

The MCL for TTHMs (0.10 mg/L) has been replaced by the final Stage 1 Disinfectants/Disinfection By-Products Rule promulgated in December 1998. The removal of total organic carbon (TOC) to reduce the formation of DBPs is achieved by the treatment technique of enhanced coagulation or enhanced softening that specifies the percentage of influent TOC that must be removed based on the raw water TOC and alkalinity levels (see Regulations).

Stage 2 Disinfectants/Disinfection By-Products Rule (Stage 2 D/DBPR)

The rule will apply to all community and nontransient, noncommunity water systems that add a disinfectant other than UV or deliver water that has been disinfected. Compliance will be based on Locational Running Annual Average (LRAA—running annual average at each sample location). Implementation of this rule will be staged. Three years after promulgation, all systems must comply with Stage 1 DBPR MCL (0.080/0.060 running annual average) and 0.120/0.100 mg/L THM/HAA5 LRAA. Six years after promulgation, large and medium systems (population served ≥10,000) must comply with 0.080/0.060 LRAA based on new sampling sites identified by Lifetime Distribution System Evaluation (LDSE). Small systems must comply with 0.080/0.060 LRAA based on new sampling sites identified by LDSE 10 years after promulgation. Final rule is anticipated by June 2003. Effective date would be June 2006.

Long-Term 1 Enhanced Surface Water Treatment Rule

The Long-Term 1 Enhanced Surface Water Treatment Rule (LT1ESWTR) applies to surface water systems (and GWUDI systems) serving less than 10,000 people. LT1ESWTR is based on the requirements for systems serving more than 10,000 people that are contained in the Interim Enhanced Surface Water Treatment Rule (IESWTR).

LT1ESWTR includes the following requirements:

1. Systems are required to achieve a 2-log removal (99 percent) of *Cryptosporidium*.

2. Systems get credit for 2-log removal of *Cryptosporidium* by meeting lower turbidity (combined filter effluent) performance standard of 0.3 ntu in 95 percent of monthly measurements, never to exceed 1 ntu (for systems using conventional or direct filtration).

3. Systems must monitor individual filter effluent and based on turbidity levels may be required to perform followup activities.

4. Systems may be required by the state to compile a disinfection profile based on the levels of DBPs in their system (80 percent of the Stage 1 MCLs is the criteria).

5. Systems looking to make a significant change in their disinfection practice will need to determine the disinfection benchmark and present that in discussions with state primacy agencies.

6. New finished-water reservoirs will need to be covered.

Long-Term 2 Enhanced Surface Water Treatment Rule (LT2ESWTR)

The rule applies to all public water systems that use surface water or GWUDI. Beginning 6 months after promulgation, monthly monitoring of the source for *Cryptosporidium* for 2 years will be used to characterize the source water. For small systems, *E. coli* concentrations will be used in lieu of *Cryptosporidium*. Based on the concentrations found, the levels of additional removal beyond that provided by conventional treatment or direct filtration will be identified. To meet the additional removal requirements, the water system would select the appropriate tool(s) from the microbial toolbox. Note that UV light disinfection would be allowed under this rule. For implementation, large and medium systems (population served $\geq 10,000$) have 3 years after source water characterization or 6 years after rule promulgation to meet the higher removal requirements. Small systems are on a delayed schedule. Final rule is anticipated by June 2003. Effective date would be June 2006.

Ground Water Rule (GWR)

GWR applies to all groundwater systems. Under this rule, primacy agencies must complete a sanitary survey that identifies any significant deficiencies that could cause contamination of the water used by consumers. A hydrogeologic sensitivity assessment must also be completed to determine the susceptibility of the groundwater source to contamination. Microbial monitoring will be required for those systems that do not disinfect, draw from a susceptible source, or have detected fecal indicators during routine monitoring. For those systems that do disinfect, have any uncorrected significant deficiencies, or draw from a susceptible source, 4-log inactivation of viruses must be demonstrated by providing adequate $C \times T$. Statutory deadline was July 2003. Effective date would be July 2005.

Radon Rule

The rule applies to all community water systems that use groundwater or mix ground and surface water. An alternative MCL of 4,000 pCi/L is allowable if the community water system or state develops a multimedia mitigation program for radon. Rule was proposed in November 1999. Final rule is expected by late 2002. Effective date would be late June 2006.

SELECTED SUPPLEMENTARY READINGS

If access to the Internet is available, the Environmental Protection Agency Web site has the latest updates on regulatory issues. Also, state Department of Health Web sites have the latest state regulatory information.

The *Federal Register* is available on the Internet and in most libraries. The *Federal Register* publishes the entire USEPA document and provides the detailed information necessary to understand a rule or act.

If access to the Internet is not available, a public or university library will have either access to the Internet or to the *Federal Register*. If a synopsis of the proposed rules and regulations and insight to upcoming legislation is desired, The *Journal of the American Water Works Association* is an excellent resource, as is the Water Quality Association.

Back to Basics Guide to Safe Drinking Water. 1990. Denver, Colo.: American Water Works Association.

Drinking Water Handbook for Public Officials. 1993. Denver, Colo.: American Water Works Association.

Guidance Manual for Compliance With the Filtration and Disinfection Requirements for Public Water Systems Using Surface Water Sources. 1990. US Environmental Protection Agency Manual. Denver, Colo.: American Water Works Association.

Introduction to Water Supply Analyses. 1982. Denver, Colo.: American Water Works Association.

Lead and Copper. 1991. Denver, Colo.: American Water Works Association.

Lead and Copper Rule Guidance Manual. Vol. 1—Monitoring. 1991. Denver, Colo.: American Water Works Association and US Environmental Protection Agency.

Lead and Copper Rule Guidance Manual. Vol. 2—Corrosion Control Treatment. 1992. Denver, Colo.: American Water Works Association and US Environmental Protection Agency.

New Dimensions in Safe Drinking Water. 1988. Denver, Colo.: American Water Works Association.

Pocket Guide to Water Sampling—Lead and Copper. 1992. Denver, Colo.: American Water Works Association.

SDWA Advisor: Regulatory Update Service. (updated regularly). Denver, Colo.: American Water Works Association.

Water Quality and Treatment. 5th ed. 1999. New York: McGraw-Hill and American Water Works Association (available from AWWA).

Water Quality Association	http://wqa.org
Federal Register	http://www.archives.gov/federal_register
<u>Rule</u>	<u>**EPA Web site**</u>
Arsenic Rule	http://www.epa.gov/safewater/arsenic.html
CCL	http://www.epa.gov/safewater/ccl/cclfs.html
Consumer Confidence Report Rule	http://www.epa.gov/safewater/ccr1.html
CROMERRR	http://www.epa.gov/cdx/cromerrr/propose/index.html
Filter Backwash Recycling Rule	http://www.epa.gov/safewater/filterbackwash.html
Ground Water Rule	http://www.epa.gov/safewater/gwr.html
IESWTR	http://www.epa.gov/safewater/mdbp/ieswtr.html
LT1ESWTR	http://www.epa.gov/safewater/mdbp/lt1eswtr.html
LT2ESWTR	http://www.epa.gov/safewater/mdbp/mdbp.html#it2
MTBE	http://www.epa.gov/safewater/mtbe.html
Public Notification Rule	http://www.epa.gov/safewater/pn.html
Radionuclides Rule	http://www.epa.gov/safewater/standard/pp/radnucpp/html
Radon Rule	http://www.epa.gov/safewater/radon.html
Six-Year Review	http://www.epa.gov/safewater/review.html
Stage 1 DBPR	http://www.epa.gov/safewater/mdbp/dbp1.html
Stage 2 DBPR	http://www.epa.gov/safewater/mdbp/mdbp.html#it2
UCMR	http://www.epa.gov/safewater/standard/ucmr/main.html

Water Quality Monitoring

All public water systems monitor water quality to some extent. Small systems with good-quality water from deep wells may only need to provide occasional monitoring. Systems using surface water sources test water quality frequently.

Water quality is monitored to meet federal and state requirements and for process control.

The contaminants that are monitored under US Environmental Protection Agency (USEPA) requirements are extensive, and public water systems must monitor water quality to ensure proper and economic treatment.

Systems using surface water sources test water quality frequently.

SAMPLING

Importance of Sampling

Sampling is a vital part of monitoring the quality of water in a water treatment process, distribution system, or source of water supply. However, errors occur easily when recording water quality information. Every precaution must be taken to ensure that the sample collected is as representative of the water source or process being examined as is feasible.

Water treatment decisions based on incorrect data may be made if sampling is not correctly performed. Representative analytical results depend on the water treatment plant operator ensuring that

- the sample is representative of the water source under consideration;

- the proper sampling techniques are used;
- the samples are protected and preserved until they are analyzed.

Types of samples

Waterworks operators collect grab samples and composite samples.

Grab samples

A grab sample is a single water sample collected at any time. Grab samples will show the water characteristics at the time the sample was taken. A grab sample may be preferred over a composite sample when

- the water to be sampled does not flow on a continuous basis;
- the water's characteristics are relatively constant;
- the water is to be analyzed for water quality indicators that may change with time, such as dissolved gases, coliform bacteria, residual chlorine, temperature, and pH.

Figures 2–1 and 2–2 illustrate this point. Figure 2–1 shows the changes in surface water dissolved oxygen (DO) over a 24-hour period. A grab sample represents the DO level only at the time the sample was taken. DO can change rapidly, e.g., the growth of algae or plants in the water.

Figure 2–2 shows the total dissolved solids (TDS) levels in the same water change very little. A grab sample accurately represents water quality for perhaps a month. TDS is a function of the minerals dissolved from rocks and soil as the water passes over or through it and may change only in relation to seasonal runoff patterns.

Composite samples

In many processes, water quality changes with time. A continuous sampler–analyzer provides the most accurate results in these cases. Many times operators are the sampler–analyzer, and continuous analysis could prove costly. Except for tests that cannot wait because of rapid physical, chemical, or biological changes of the sample (such as tests for DO, pH, and temperature), a fair compromise may be reached by taking samples throughout the day at hourly or 2-hour intervals. Each sample should be refrigerated immediately after collection. At the end of 24 hours, each sample is vigorously mixed and a portion of each sample is then withdrawn and mixed with the other samples. The size of the portion is in direct proportion to the flow when the sample was collected and the total size of sample needed for testing. For example, if hourly samples were collected when the flow is 1.2 mgd, use a 12-mL portion sample, and when the flow is 1.5 mgd, use a 15-mL portion of the sample. The resulting mixture of

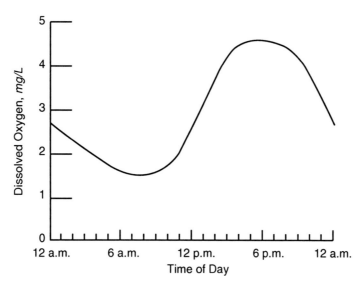

FIGURE 2–1 Example of hourly changes in dissolved oxygen for a surface water source

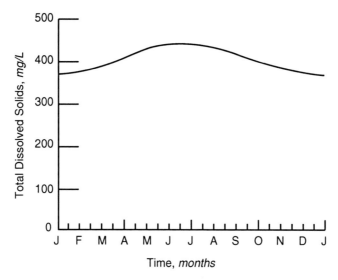

FIGURE 2–2 Example of monthly changes in total dissolved solids for the surface water source shown in Figure 2–1

portions of samples is called a *composite sample*. **In no instance should a composite sample be collected for bacteriological examination.**

When the samples are taken, they can either be set aside or combined as they are collected. In both cases, they should be stored at a temperature of 104°F (40°C) until they are analyzed.

Sampling Point Selection

Careful selection of representative sample points is an important step in developing a sampling procedure that will accurately reflect water quality. The criteria used to select a sample point depend on the type of water sampled and the purpose of the testing. Samples are generally collected from three broad types of areas:

- Raw-water supply
- Treatment plant
- Distribution system

Raw-water sample points

The choice of collection points for raw-water samples depends on the type of system being sampled. There are at least three general types of systems:

- Raw-water transmission lines
- Groundwater (wells)
- Rivers, reservoirs, and lakes

Raw-water transmission lines and groundwater sources are sampled directly from the transmission line or well-discharge pipe. After a sampling point has been selected, the pipeline is equipped with a small sample valve, often called a *sample cock* (Figure 2–3). The valve must be fully opened before sampling to flush out any standing water and accumulated sediment.

Most of the physical factors known to promote mixing in surface waters are absent or are much less effective in groundwater systems. Wells usually draw water from a considerable thickness of saturated rock and often from several different strata. These water components are mixed by the turbulent flow of water in the well before they reach the surface and become available for sampling. Most techniques for well sampling and exploration are usable only in unfinished or non-operating wells. Usually the only means of sampling the water tapped by a well is a collection of a pumped sample. The operator is cautioned to remember that well pumps and casings can contribute to sample contamination. If a pump has not run for an extended period of time prior to sampling, the water collected may not be representative of the normal water quality.

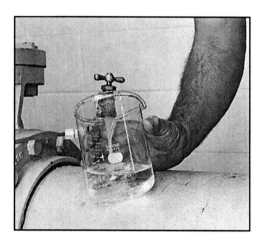

FIGURE 2–3 Sample cock attached to pipeline for sampling

Rivers. To adequately determine the composition of a flowing stream, each sample (or set of samples taken at the same time) must be representative of the entire flow at the sampling point at that instant. The sampling process must be repeated at a frequency sufficient to show changes of water quality that may occur over time in the water passing the sampling point.

On small or medium-sized streams, it is usually possible to find a sampling point at which the composition of the water is presumably uniform at all depths and across the stream. Obtaining representative samples in these streams is relatively simple. For larger streams, more than one sample may be required. A portable conductivity meter is very useful in selecting good sample sites.

Reservoirs and lakes. Water stored in reservoirs and lakes is usually poorly mixed. Thermal stratification and associated depth changes in water composition (such as DO) are among the most frequently observed effects. Single samples can therefore be assumed to represent only the spot of water from which the sample came. Therefore, a number of samples must be collected at different depths and from different areas of the impoundment to accurately sample reservoirs and lakes. See Figures 2–4 and 2–5.

Treatment plant sample points
Treatment plants are sampled to evaluate the treatment efficiency of unit processes or to evaluate operational changes. Selection of in-plant sample points is an important step in developing an overall process-control program for a water

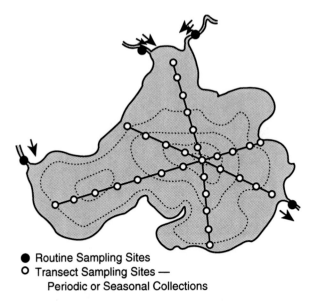

● Routine Sampling Sites
○ Transect Sampling Sites —
    Periodic or Seasonal Collections

Source: *Mackenthun and Ingram (1967).*

FIGURE 2–4 Routine and transect sample points in a natural lake

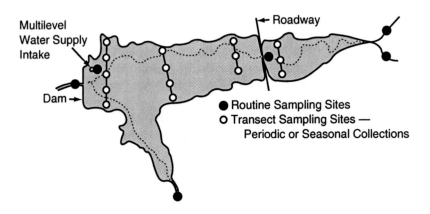

Source: *Mackenthun and Ingram (1967).*

FIGURE 2–5 Routine and transect sample points in a reservoir

treatment plant. Samples from the points selected can be tested to determine the efficiency of the various treatment processes. The test results will also help to indicate operational changes that will improve contaminant removal efficiencies or reduce operating costs.

Collection of representative samples within the water treatment plant is similar to sample collection in a stream or river. The operator must ensure that the water sampled is representative of the water passing that sample point. In many water plants, money is spent to purchase sample pumps and piping only to sample from a point that is not representative of the passing water. A sample tap in a dead area of a reservoir or on the floor of a process basin serves no purpose in helping the plant operator with control of water quality. The operator is urged to find each and every sample point and ensure that it is located to provide a useful and representative sample. If the sampling point is not properly located, the operator should make arrangements to relocate the piping to a better location.

Treatment plants vary widely in the kinds of treatment processes used and the configurations of the processes. In general, in-plant sample points are established at every place where, because of a treatment method or group of methods, a measurable change is expected in the treated-water quality. Figure 2–6 identifies 10 suggested locations where process-control samples are routinely collected in a plant employing several different treatment processes. These locations are described in the following list.

- Between sample points 1 and 2, test results should show a reduction in algae and the associated tastes and odors (the result of chemical pretreatment), a reduction in sediment load (the result of presedimentation), and a reduction in debris (the result of screening).

- Between points 2 and 3, aeration should cause a significant reduction in dissolved gases and in oxidation of iron and manganese.

- Between points 3 and 4, the combined effects of coagulation, flocculation, and sedimentation should cause a reduction in turbidity and color.

- Water quality changes between sample points 4 and 5 will allow the operator to monitor the effectiveness of the softening process.

- Sample points 5 and 6 allow monitoring of the efficiency of filtration in removing turbidity and previously oxidized iron and manganese as well as the reduction in pathogenic organisms.

- Sampling at points 6 and 7 will indicate the efficiency of the adsorption process in removing organic chemicals.

- Point 8 is used for the measurement of fluoride concentration to ensure that water entering the distribution system contains the proper level.

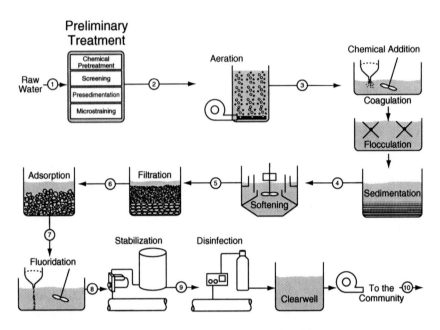

FIGURE 2–6 Suggested in-plant sample points (indicated by numbered circles)

- Sampling at point 9 will provide a final check on pH and alkalinity for corrosion control.
- Point 10 is used for monitoring chlorine residual, turbidity, and the presence of coliform bacteria in the finished water.

In the selection of in-plant sample collection points, certain precautions should be kept in mind. Points immediately downstream from chemical additions should be avoided because proper mixing and reaction may not have had time to take place. Samples should always be taken from the main stream of flow, avoiding areas of standing water, algae mats, or other floating debris.

Finished-water sample points are normally established downstream of the final treatment process at or just before the point where the water enters the distribution system, such as the point of discharge from the clearwell. For example, turbidity samples required by the National Primary Drinking Water Regulations (NPDWRs) must be collected before the water enters the distribution system.

Distribution system sample points

Representative sampling in the distribution system is a true indication of system water quality. Results of sampling should show if there are quality changes in the entire, or parts of the, system and may point to the source of a problem (such as tastes and/or odors). Sampling points should be selected, in part, to trace the course from finished water source (at the well or plant) through the transmission mains, and then through the major and minor arteries of the system. A sampling point on a major artery, or on an active main directly connected to it, would be representative of the water quality.

Sample points in the distribution system are used to determine the quality of water delivered to consumers. In some cases the distribution system samples may be of significantly different quality than samples of finished water at the point of entry to the distribution system. For example, corrosion in distribution system pipelines can cause increases in water color, turbidity, taste and odor, and physical constituents such as lead and copper. Microbiological growth can also be taking place in the water mains, which will degrade water quality. More seriously, a cross-connection between the distribution system and a source of contamination can result in chemical or microbiological contamination of the water in the system.

Most of the samples collected from the distribution system will be used to test for coliform bacteria and chlorine residual. Others may be used to determine water quality changes. Still others will be used to test for compliance with the inorganic, organic, and radiochemical maximum contaminant levels (MCLs) and with the Lead and Copper Rule, as required by the applicable drinking water standards. Distribution system sampling should always be performed at points representative of conditions within the system.

The two major considerations in determining the number and location of sampling points are that they should be

- representative of each different source of water entering the system;
- representative of conditions within the system, such as dead ends, loops, storage facilities, and pressure zones.

The precise location of sampling points depends on the configuration of the distribution system. The following examples provide some general guidance for sample point selection.

Example 1. Figure 2–7 provides an example of how sample points may be selected for a small surface water distribution system serving a population of 4,000. This is a typical small branch system having one main water line and several branch or dead-end water lines. For this system, a single point, *A*, is sufficient for turbidity sampling. This point is representative of all treated water entering the distribution system.

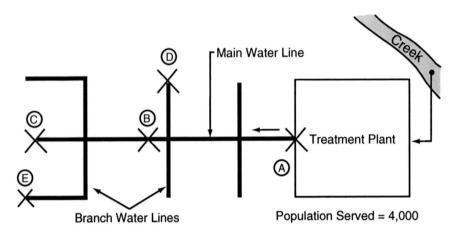

FIGURE 2–7 Sampling points (indicated by x) in a typical small-branch distribution system

The NPDWRs require a minimum of four bacteriological samples per month for a community of 4,000, to be taken at four different points in the system. Point B represents water in the main line, and point C represents water quality in the main-line dead end. Points D and E were selected to produce samples representative of a branch line and a branch-line dead end, respectively.

Consideration of how often and at what times these points are sampled is also necessary to ensure that the samples accurately represent conditions in the distribution system. Although the minimum requirement of four samples per month could be met by collecting samples from all points on 1 day, this sampling frequency would not produce samples that represented bacteriological conditions within the system throughout the month. A better program would be to sample points B and E at the beginning of the month and points C and D at mid-month. Sampling should be representative both in location and in time.

Although this type of program is adequate to meet the minimum monitoring requirements of the NPDWRs, good operating practices would include periodic sampling at each dead end and several additional sampling points within the distribution system, with samples taken each week. The exact number and location of these operational sampling points depend on the characteristics of the specific system and on state requirements.

Chlorine residual samples should be taken from each sample point when bacteriological samples are collected and should be analyzed as soon as possible. Sampling for routine water chemistry, along with the required sampling for inorganic and organic chemicals and radionuclides, also can be conducted at one of the coliform sampling points.

Sampling for a similar system using a groundwater source would be the same, except that turbidity sampling is generally not required and samples for organic chemical analysis must be collected at each well.

Example 2. Figure 2–8 illustrates a typical small-loop distribution system having one main loop and several branch loops, serving a population of 4,000. One turbidity sample point, *A*, is sufficient because that point is representative of all treated water entering the distribution system.

For bacteriological sampling, two sampling points, *B* and *C*, are adequate. Point *B* is representative of water in the main-line loop, and point *C* is representative of water in one of the branch-line loops. To produce the required minimum of four samples per month, points *B* and *C* can be sampled on alternate weeks, or additional similar sampling points can be selected. However, good operating practice would include two to three times this number of samples, depending on the characteristics of the particular system. As with the system in the previous example, chlorine residual samples should be taken whenever bacteriological sampling is performed.

Example 3. Figure 2–9 illustrates a system serving a population of 17,440 that obtains water from both a creek and a well. The distribution system has the features of both the branch and the loop systems shown in Figures 2–7 and 2–8.

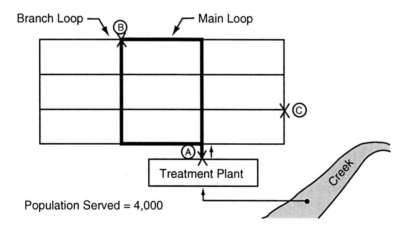

FIGURE 2–8 Sampling points (indicated by x) in a typical small-loop distribution system

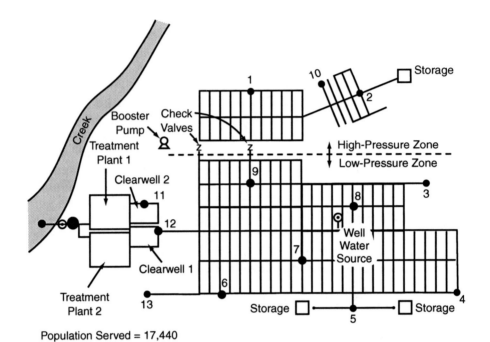

Population Served = 17,440

FIGURE 2-9 Sampling points (indicated by solid dots) in a medium-size system with surface and groundwater sources

To determine sample-point locations, the following four questions should be considered:

1. What tests must be run?
2. From what locations will the samples be collected?
3. How often must the samples be taken?
4. How many sampling points will be needed?

The answers to the first and third questions—what tests must be run and how often—may vary from state to state, and they are also likely to change periodically in response to changes in federal requirements.

Additional samples may also be required for the system's own quality control (QC) program. Examples include taste and odor, color, pH, TDS, iron, manganese, and heterotrophic plate count.

Once the tests and test frequencies have been determined, the number and specific locations of sampling points must be selected. The NPDWRs require a turbidity sample to be taken at each point where surface water enters the distribution system. Because waters from parallel treatment plants enter two separate clearwells in Figure 2–9, two turbidity sampling points are required (points 11 and 12). The well will not have to be sampled for turbidity, but periodic sampling directly from the well for chemical quality analysis will be required as directed by the state.

In the selection of sample points that will be representative for coliform analysis, a variety of factors must be considered:

- Uniform distribution of the sample points throughout the system;
- Location of sample points in both loops and branches;
- Adequate representation of sample points within each pressure zone;
- Location of points so that water coming from storage tanks can be sampled;
- For systems with more than one water source, location of sample points in relative proportion to the number of people served by each source.

On the basis of these fundamental considerations, bacteriological sample points can be selected. A treatment plant serving a community with a population of 17,440 must test 20 coliform bacteria samples per month, according to the NPDWRs. After a careful review of the configuration of the distribution system layout, 10 coliform bacteria sample sites were selected. The reasons for the selection of each point shown in Figure 2–9 are as follows:

- Point 1 is on the main loop in the high-pressure zone; it should produce representative samples for that part of the system.
- Point 2 is on the branch loop in the high-pressure zone, representative of storage flow to the system.
- Point 3 is on a dead end. Some authorities advise against dead-end sampling points because they do not produce representative samples. However, consumers do take water from branch-line dead ends. In the case of the example, there are seven branch-line dead ends that no doubt serve significant numbers of consumers. It is representative to have one or two sample points on these branch lines at or near the end. If there are indications of bacteriological problems in sampling branch-line dead ends, hydrants and blowoff valves should be flushed and branch lines resampled immediately to determine if the problem has been corrected. If the problem persists, additional investigation is needed to locate the source of contamination.

- Point 4 is located on the main loop of the low-pressure zone and represents water from treatment plant 2, the well water source, the storage tanks, or any combination (depending on system demand at sampling time).
- Point 5 allows for sampling of the quality of water flowing into the system from storage.
- Points 6 through 9 were selected by uniformly distributing points in the low-pressure zone, the zone that serves the major part of the community.
- Point 10 was selected as representative of a branch-line dead end in the high-pressure zone, just as point 3 was selected in the low-pressure zone.
- Points 11 and 12, as stated previously, are used as turbidity monitoring points.
- Point 13 was added to monitor a dead-end branch that is fairly isolated from other sampling points yet serves a large population.

Sample faucets

Once representative sample points have been located on the distribution system map, specific sample faucets must be selected. In many cases, suitable faucets can be found inside public buildings, such as fire stations or school buildings, inside the homes of water system or municipal employees, or inside the homes of other consumers.

In smaller water systems, special sample taps are not available. Therefore, customers' faucets must be used to collect samples. Indoor taps are best, if available. Front-yard faucets on homes supplied by short service lines (homes with short service lines are located on the same side of the street as the water main) will suffice if there are no other options. Dipping them in bleach first is one way to assure that the tap will not taint the sample.

If the customer is home, contact the person in the home and obtain permission to collect the sample. Disconnect the hose from the faucet if one is attached and don't forget to reconnect the hose when finished collecting the sample. Open the faucet to a convenient flow for sampling (usually about a half-gallon per minute). Allow the water to flow until the water in the service line has been replaced twice. Because 50 ft (15 m) of three-quarter-inch (18-mm) pipe contains over 1 gal (3.8 L), 4 or 5 minutes will be required to replace the water in the line twice. Collect the sample. Be sure the sample container does not touch the faucet.

Do not try to save time by turning the faucet handle to wide open to flush the service line. This will disturb sediment and incrustations in the line that must be flushed out before the sample can be collected.

For sampling, it is also best to try to find a faucet that does not have an aerator. If a faucet with an aerator must be used, follow the state primacy agency's recommendation on whether the aerator should be removed for sample collection.

Once a representative sample point has been selected, it should be described on the sample record form so that it can be easily located for future sample collection.

Collection of Samples

The steps described in the following sections represent general sample collection procedures that should be followed regardless of the constituent tested. Special sample-collection procedures required for certain tests are described in succeeding chapters.

Only containers designed for water sampling and provided by the laboratory should be used. Mason jars and other recycled containers cannot be trusted to function properly no matter how well they are cleaned and are generally not accepted by a laboratory for water analysis. Some laboratories reuse sample containers by washing them under carefully controlled conditions. In other cases, it has been found more economical to dispose of used bottles and provide only new ones for collection.

When a container with a screw-on lid is used, the lid should be removed and held threads down while the sample is collected in the container. The lid can easily be contaminated if the inside is touched or if the lid is set face down or placed in a pocket. A contaminated lid can contaminate the sample, which will necessitate resampling, causing a great deal of unnecessary time and expense.

Raw-water sample collection

A clean wide-mouth sampling bottle should be used for raw-water sampling. The bottle should not be rinsed; this is especially important if the bottle has been pretreated or contains a preservative. The open bottle should be held near its base and plunged neck downward below the surface of the water. The bottle should then be turned until the neck points slightly upward for sampling, with the mouth directed toward any current present. Care must be taken to avoid floating debris and sediment. In a water body with no current, the bottle can be scooped forward to create a current. Once the bottle has been filled, it is retrieved, capped, and labeled.

If the sampler is wading, the sample bottle should be submerged upstream from that person. If a boat is being used for stream sampling, the sample should be taken on the upstream side.

When samples are being taken from a large boat or a bridge, the sample bottle should be placed in a weighted frame that holds it securely. The opened bottle and holder are then slowly lowered toward the water with a small-diameter rope. When the bottle approaches the surface, the unit is dropped quickly into the water. Slack should not be allowed in the rope because the bottle could hit bottom and break or pick up mud and silt. After the bottle is filled, it is pulled in, capped, and labeled.

Treatment plant sample collection

The sample-collection procedure used to take samples from an open tank or basin or in an open channel of moving water is essentially the same as for raw-water sampling. Treatment plants are equipped with sample taps. These faucets provide a continuous flow of water from various locations in the treatment plant. To collect a sample, the operator or laboratory technician draws the required volume from the sample tap. Figure 2–10 shows a typical bank of sample faucets in a laboratory.

Distribution system sample collection

Once the distribution system sample locations have been selected, sample collection consists of a few simple, carefully performed steps. First, the faucet is turned on and set to produce a steady, moderate flow of water (Figure 2–11). If a steady flow cannot be obtained, the tap should not be used. The water is allowed to run long enough to flush any stagnant water from the service line. This process usually takes 2 to 5 minutes. The line is usually clear when the water temperature drops and stabilizes. The sample is then collected without the flow changing. The sample bottle lid should be held threads down during sample collection and replaced on the bottle immediately. The exception to this procedure is with samples collected for lead and copper analyses; these must be first-draw samples, collected immediately after the faucet has been opened. The final step after sampling is to label the bottle.

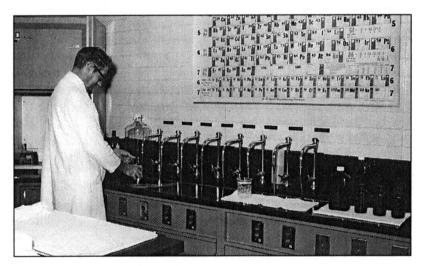

FIGURE 2–10 Sample faucets in a laboratory

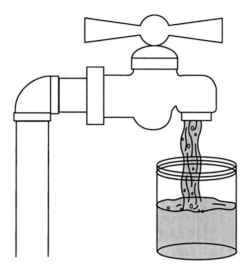

FIGURE 2-11 Sample faucet should be set to produce a steady, moderate flow

A once common practice was to flame the outside of a faucet with a propane torch. This procedure is no longer recommended. Experience showed that the flame could not be held on the faucet long enough to kill all the bacteria on the outside of the faucet without potential damage to the faucet. Many faucets are now made partly or entirely of plastic, which will quickly melt if high heat is applied. However, dipping the faucet in a small bowl of bleach is still a common practice.

Samples should not be collected from sill cocks or other faucets with hose threads. Because of the way they are constructed, these faucets will not usually throttle to a smooth flow. Also if any water splashes up onto the threads and then drains into the sample bottle, it will bring with it contaminants from the outside of the faucet.

Special-purpose samples

Occasionally a water utility may need to collect samples for special testing purposes. Procedures in such cases depend on the reason for the sampling.

For example, a consumer may have complained about taste, odor, or color in the water. In such a case, samples are collected from the consumer's faucet to determine the source of the problem. The faucet is opened and a sample taken immediately. This sample represents the quality of water standing in the service line. The water is then allowed to run for 2 to 5 minutes or until the water temperature changes, so that the standing water in the service line is completely flushed out; then a second sample is taken. The second sample is fresh from the distribution system. Comparing test results from the two samples will often help identify the origin of the problem causing the consumer complaint.

Customer complaints of taste, odor, or color are often caused because the consumer's hot-water heater, water softener, or home water-treatment device is not operating properly. If the hot-water supply is suspected, the first sample should be collected from the hot-water tap. The tap is turned on and allowed to run until the water is hot before the sample is collected. A second sample representing the water in the service line should be taken from the cold-water tap as previously described. Comparing the test results from the two samples will help identify the origin of the problem.

There are many other reasons for taking special-purpose samples. The above example emphasizes the importance of knowing for what the sample test results will be used so that the sample collected will be representative of the conditions tested.

MONITORING FOR CHEMICAL CONTAMINANTS

Drinking water may contain contaminants considered a threat to the public. The contaminants of concern may occur naturally in the water, be human-made, or be formed in the water treatment process. The chemicals are broken into four general classes for regulation:

1. Inorganic chemicals (IOCs)
2. Synthetic organic chemicals (SOCs)
3. Volatile organic chemicals (VOCs)
4. Radionuclides (covered in chapter 8 of this book)

Monitoring Requirements

The need to establish regulations for new chemical contaminants has presented USEPA with the problem of *Analytical techniques.* Before a requirement to monitor for a contaminant can be imposed, the testing methods must be developed to ensure that an adequate number of laboratories will be available to perform the tests and that they will get consistent, reliable results. Many of the chemicals now being added to the regulated contaminant list must be analyzed at the parts-per-billion level or in even smaller concentrations.

Faucets selected should be on the lines connected directly to the main. Only cold-water faucets should be used for sample collection. A sampling faucet must not be located too close to a sink bottom. Contaminated water or soil may be present on the exteriors of such faucets, and it is difficult to place a collection bottle beneath them without touching the neck interior against the faucet's outside surface. Samples should not be taken from the following types of faucets (Figure 2–12):

- Leaking faucets, which allow water to flow out around the stem of the valve and down the outside of the faucet.
- Faucets with threads.
- Faucets connected to home water-treatment units, including water softeners.

LABORATORY CERTIFICATION

Each of the approximately 200,000 public water systems affected by the Safe Drinking Water Act (SDWA) must routinely monitor water quality to determine if it is adequately protected from regulated microbiological, chemical, and radiological contaminants. It is imperative that the analyses for all of this monitoring be performed by standard methods that will yield consistent results. Consequently, states are required by federal regulations to consider analytical results from water systems only if the water samples have been analyzed by a certified laboratory. The only exceptions are measurements for turbidity, chlorine residual, temperature, and pH, which may be performed by a person acceptable to the state using approved equipment and methods.

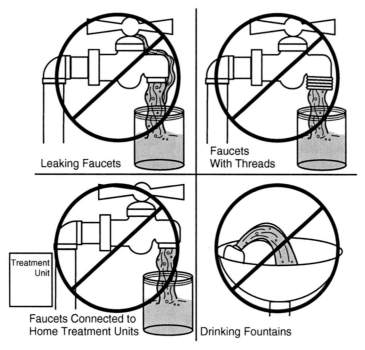

FIGURE 2-12 Types of faucets that should *not* be used for sampling

The federal regulations require each state with primary enforcement responsibility to have available laboratory facilities that have been certified by USEPA, with capacities sufficient to process samples for water systems throughout the state. Certified laboratories fall into the following general classes:

- State-operated laboratories
- Water system laboratories
- Commercial laboratories

In most states, the necessary capacity is provided by all three types of laboratories. Some laboratories may be certified to perform only one type of analysis; for instance, some laboratories are set up to handle only microbiological analyses. Analyses requiring expensive equipment and highly trained technicians, such as for organic chemical and radiological monitoring, are also generally handled by specialized laboratories.

Consistency among laboratories in analytical results is maintained by a USEPA program that periodically provides to each laboratory carefully prepared samples containing a known concentration of a contaminant for each of the types of analyses for which the laboratory is certified. The values of the samples are unknown to the laboratory, and its staff must be able to determine the contaminant concentration within an appropriate tolerance in order to maintain the laboratory's certification.

Historically, most states have operated their own laboratories to process water system samples. But the number of samples has increased severalfold in recent years, so it is difficult for the states to continue providing laboratory service with state funding only. Some states have instituted charges to water systems to help fund the laboratory services. Other states will only process a certain number of samples from any one water system, and if more are required, commercial laboratories must be used.

RECORD KEEPING AND SAMPLE LABELING

Records should be kept for every sample that is collected. A sample identification label or tag should be filled out at the time of collection. Each label or tag should include at least the following information:

- Water utility name
- Water system's Public Water System Identification number
- Date sample was collected
- Time sample was collected
- Location where sample was collected

- Type of sample—grab or composite
- Tests to be run
- Name of person sampling
- Preservatives used
- Bottle number

The samples provided to laboratories should always be clearly labeled. The information on the label should also be entered on a record-keeping form that is maintained as a permanent part of the water system's records.

SAMPLE PRESERVATION, STORAGE, AND TRANSPORTATION

Samples cannot always be tested immediately after they are taken. Ensuring that the level of the constituent remains unchanged until testing is performed requires careful attention to techniques of sample preservation, storage, and transportation. It is also extremely important that records be kept of the chain of custody of samples collected for SDWA compliance.

Preservation and storage

Once a sample has been collected, its quality begins to change because of chemical and biological activity in the water. Some characteristics, such as alkalinity, pH, dissolved gases, and odor, can change quickly and quite significantly; therefore, samples to be analyzed for these parameters must not be stored under any conditions. Other parameters, such as pesticides and radium, change more slowly and much less noticeably, and these samples can usually be stored for considerable lengths of time if necessary.

To extend the storage time of samples for chemical analysis, sample-preservation techniques have been developed that slow the chemical or biological activity in the sample. This allows it to be transported to the laboratory and tested before significant changes occur.

Sample preservation usually involves one of two basic steps:

- Refrigeration
- pH adjustment

The storage time of some samples can be prolonged by keeping them refrigerated until the analysis is performed. In some cases, it is recommended that samples be transported or shipped to the laboratory in a portable cooler containing an ice pack.

In many cases, the laboratory provides bottles for specific analyses with the preservative already added. It is particularly important not to allow these containers to overflow as they are filled, or some of the preservative will be lost. These containers must also be kept out of the reach of children because the preservative material could be harmful to a child who opens a container. If preservatives are to be added by the sampler, specific instructions on the procedures should be obtained from the laboratory that will perform the analyses.

Time of sampling

Most laboratories do not maintain a full staff on weekends, so they generally request that samples that must be processed immediately upon arrival, such as bacteriological samples, be collected and shipped early in the week. If a sample arrives on a weekend and cannot be processed, the delay will probably exceed the required holding time, and the sample will have to be rejected. However, most laboratories will accept emergency samples on weekends.

Samples that must be submitted within a specified compliance period should generally be collected and sent to the laboratory early in the compliance period. Some of the problems that can require resampling are described in the following list.

- The sample is frozen or broken during shipment.
- The sample is lost or delayed in shipment and arrives at the laboratory after the specified holding time has elapsed.
- The laboratory makes an error in processing the sample.
- The laboratory analysis is inconsistent or exceeds the MCL, and another sample to confirm the results is required.

Sampling early in the compliance period ensures that time is available for one or more resamplings, if necessary, before the end of the period. If resampling has not been completed before the end of a compliance period, a water system is usually deemed out of compliance and will be instructed by the state to provide public notification.

Transportation

If samples arrive at a laboratory past the specified holding time following collection, the laboratory must reject the samples. New sample bottles must then be shipped to the water system and another set of samples will have to be collected and shipped back.

The US mail is usually the best and easiest method of shipment. If regular mail service fails to deliver samples reliably within the required time period, Express

Mail or package delivery services may be tried. In some cases, changing to a laboratory at a different location may improve delivery time. Some water system operators who are located near a laboratory have found it best just to drive the samples directly to the laboratory.

If samples are shipped, it is important to make sure that the bottle caps are tight to prevent leakage. Some systems that have had bottle caps loosen during shipment have found that wrapping the lids with electrical tape is an easy method of further securing them. Samples must be packed in a sturdy container with enough cushioning material to prevent breakage. The box should be marked to indicate which end is up, that the contents are fragile, that they must not be allowed to freeze, and that priority should be given to the shipment.

Chain of Custody

As more and more parameters are added to the list of regulated and unregulated contaminants, and as the MCLs and MCLGs are set closer to zero, the practice of good quality assurance and quality control (QA/QC) procedures becomes very important. One essential part of QA/QC is maintaining a written record of the history of SDWA compliance samples from the time of collection to the time of analysis and subsequent disposal. This record, called the *chain of custody*, is important if the analyses are ever challenged and need to be defended. Chain-of-custody requirements vary by state, so water system operators should be sure that the requirements for their state are being met.

Field log sheet

One method of establishing the chain-of-custody record is to use a daily field log sheet, which should contain the following information:

- Date the samples were collected
- Name of the sampler
- List of all the samples collected by the sampler on this date
- List of all the sample locations for this date
- Time of day each sample was collected
- Comments concerning any unusual situations
- Signature of the individual receiving the samples from the sampler
- Date and time the samples were received by the laboratory
- Location or identification of the laboratory

This log sheet states that the samples were in the custody of the sampler until they were turned over to the shipper. The laboratory record then follows the history of the sample to disposal.

Sampler's liability

If the results of an analysis of a specific sample are ever questioned, the sampler will be asked to verify that the sample was in his or her custody until it was turned over or sent to the laboratory. The sampler will be asked to verify that the sample was collected, stored, and transported using proper procedures and that no other person could have in any way altered the concentrations of any contaminant(s) present.

Sampler's responsibility

The sampler has the basic responsibility to assure that the sample is collected, labeled, stored, and transported to the laboratory in the proper manner. The sample collector must also be able to testify that the sample was under his or her custody at all times.

SELECTED SUPPLEMENTARY READINGS

Carey, E. 1992. Responsible Sampler, Lab's Best Asset. *Opflow*, 18(8):5.

————. 1992. Water Quality Only as Good as the Sample. *Opflow*, 18(8):5.

Eaton, A., G. Lynch, and K. Thompson. Getting the Most From a Contract Laboratory. *Jour. AWWA*, 85(9):44.

Feige, M.A., C. Madding, and E.M. Glick. 1993. USEPA's Drinking Water Laboratory Certification Program. *Jour. AWWA*, 85(9):63.

Lay, T. 1989. Proper Sampling Helps Systems Comply With SDWA. *Opflow*, 15(1):3.

Lee, B.H., R.A. Deininger, and R.M. Clark. 1991. Locating Monitoring Stations in Water Distribution Systems. *Jour. AWWA*, 83(7):60.

Lisle, J. 1944. *An Operator's Guide to Bacteriological Testing.* Denver, Colo.: American Water Works Association.

Mackenthun, K.M., and W.M. Ingram. 1967. *Biological Associated Problems in Freshwater Environments.* Cincinnati, Ohio: US Department of the Interior, Federal Water Pollution Control Administration.

Pocket Guide to Water Sampling—Inorganic Contamination. 1991. Denver, Colo.: American Water Works Association.

Pocket Guide to Water Sampling—Lead and Copper. 1992. Denver, Colo.: American Water Works Association.

Pocket Guide to Water Sampling—Microbiological Contamination. 1990. Denver, Colo.: American Water Works Association.

Pocket Guide to Water Sampling—Organic Contamination. 1990. Denver, Colo.: American Water Works Association.

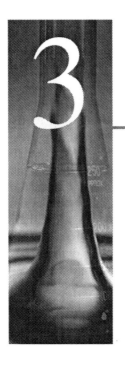

3

Water Laboratory Equipment and Instruments

The material most suitable for the majority of the bottles, beakers, and other containers used in the laboratory is heat-resistant borosilicate glass.

Water treatment processes cannot be controlled effectively unless the operator has some means to check and evaluate the quality of water being treated and produced. Laboratory quality control tests provide the necessary information to monitor the treatment processes and ensure a safe and good-tasting drinking water for all who use it. By relating laboratory results to treatment operations, the water treatment or supply system operator can first select the most effective operational procedures, then determine the efficiency of the treatment processes, and identify potential problems before they affect finished water quality. For these reasons, a clear understanding of laboratory procedures is a must for every waterworks operator.

LABWARE

The material most suitable for the majority of the bottles, beakers, and other containers used in the laboratory is heat-resistant borosilicate glass. It is commonly sold under the trade names Pyrex® or Kimax®. This heat-resistant glass can be sterilized repeatedly at high temperature and pressure; can be heated over open flames without shattering; and can also withstand heat generated from chemical reactions. However, rapid heating and cooling can weaken even heat-resistant glass, eventually causing it to crack or shatter.

Plastic is the second most common labware material and is suitable for many laboratory purposes. Some types of plastic are resistant to high temperatures and can be autoclaved. Extensive use of plastic labware is a matter of choice. Its principal advantage is that it is unbreakable. Some types of plastic labware are also disposable, which eliminates the need for laborious cleaning procedures. However, reusable plastic is harder to clean and cannot be used for all chemical analyses. For example, plastic labware should not be used in preparing samples for organic chemical analysis because the plastic will absorb the organic compound, causing erroneous results. In certain other tests, such as extractions using organic solvents, the chemicals used may deteriorate plastic almost immediately. Additionally, plastic labware is easily scratched and marred and will eventually become cloudy with use.

Another labware material is soft (nonheat-resistant) glass, which can be used to store some dry chemical reagents, such as calcium, magnesium, sulfate, and chloride. This material is not recommended for extensive laboratory use because it breaks easily and cannot be heated.

Some of the common types of laboratory containers are described in the following sections.

Beakers

A beaker is a glass jar with an open top, vertical side walls, and a lip that simplifies pouring of liquids (Figure 3–1). Common laboratory beakers range in size from 25 to 4,000 mL. The 250- and 500-mL sizes are the most popular. Beakers are used as mixing vessels for most chemical analyses, and an ample supply of various sizes should always be kept on hand for use in a laboratory.

FIGURE 3-1 Beakers

Burettes

A burette is a glass tube, graduated over part of its length, and fitted with a stop-cock (Figure 3–2). The most common sizes are 10, 25, and 50 mL. The graduations are normally in tenths of a milliliter. Burettes are designed for measuring and dispensing solutions during titration, a procedure commonly used when determining the concentration of a substance in solution. Both glass and plastic burettes are available. Plastic burettes are especially useful for field tests. Bottletop burettes (Figure 3–2) are also available; they have the advantage of being easy to read, and allow more rapid titration.

Dilution Bottles

Dilution bottles are also known as *milk dilution bottles* or *French squares*. They are autoclavable glass or plastic vessels used for diluting bacteriological samples for analysis. The bottles are square (Figure 3–3), with narrow mouths threaded to receive a screw cap. All bottles have a 160-mL capacity with a mark at the 99-mL level to facilitate 1-to-100-mL dilutions of a sample.

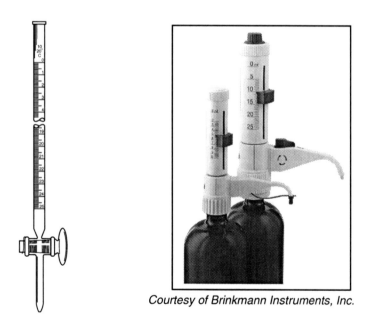

Courtesy of Brinkmann Instruments, Inc.

FIGURE 3–2 Manual burette *(left)* and bottletop burette *(right)*

FIGURE 3-3 Dilution bottle

Flasks

There are many types of flasks, each with its own specific name and use (Figure 3–4). Some names, such as *distilling* and *filtering*, identify their use. Other names specify the test they are used for, such as Kjeldahl. All flasks have narrow necks. Erlenmeyer and volumetric flasks are the most common types.

Erlenmeyer flasks

One of the most frequently used pieces of labware is the Erlenmeyer flask. Ranging in size from 100 to 4,000 mL, these flasks are characterized by their bell shape. They are recommended for mixing or heating chemicals because they minimize splashing. They are also frequently used for preparing and storing culture media.

Volumetric flasks

Volumetric flasks have long, narrow necks. They range in size from 10 to 2,000 mL; an etched ring around the neck indicates the level at which the flask's capacity is reached. Volumetric flasks are used for preparing and diluting standard solutions. Because these flasks are designed for measuring, they should not be used for long-term storage of solutions.

Funnels

The funnel is a common piece of laboratory equipment. Four of the most frequently used types are shown in Figure 3–5. The general-purpose funnel is used to transfer liquids into bottles or to hold filter paper during a filtering operation. Funnels are made of heat-resistant glass, soft glass, or plastic. There are also several disposable types.

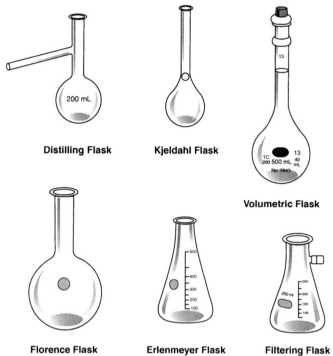

Distilling Flask   Kjeldahl Flask

Volumetric Flask

Florence Flask   Erlenmeyer Flask   Filtering Flask

FIGURE 3–4 Flasks

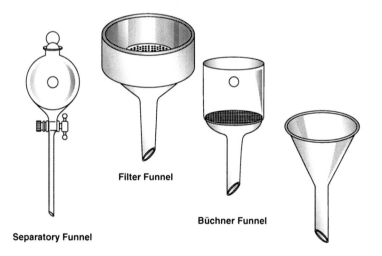

Filter Funnel

Büchner Funnel

Separatory Funnel

General-Purpose Funnel

FIGURE 3–5 Funnels

Graduated Cylinders

Graduated cylinders are tall, slender, cylindrical containers made of glass or plastic (Figure 3–6). They generally have a pour spout and a hexagonal base. They range in size from 10 to 4,000 mL. Graduations are marked in 0.2-mL intervals on the 10-mL size and in 50-mL intervals on the 4,000-mL size. Graduated cylinders are used for measuring liquids quickly but without great accuracy.

Petri Dishes

Petri dishes are shallow dishes with vertical sides and flat bottoms. They usually have loose-fitting covers (Figure 3–7). They are used as containers for culturing standard plate counts and membrane filters. They may be glass or plastic and must be completely transparent for optimum visibility of colonies.

Usually 100-mm ×15-mm dishes are used for heterotrophic plate counts (HPCs). A 50-mm × 12-mm petri dish with a tight bottom lid is used to contain and culture 47-mm membrane filters. The tight fit retards evaporation loss from both broth and agar media, which helps maintain humidity inside the dish.

Pipettes

Pipettes are used for accurate volume measurements and transfer. There are three types of pipettes commonly used in the laboratory—volumetric pipettes, graduated or Mohr pipettes, and serological pipettes.

FIGURE 3-6 Graduated cylinder   **FIGURE 3-7 Petri dish**

Volumetric pipettes are available in sizes, such as 1, 10, 25, 50, and 100 mL. They are used to deliver a single volume. Measuring and serological pipettes, however, will deliver fractions of the total volume indicated on the pipette.

In emptying volumetric pipettes, they should be held in a vertical position and the outflow should be unrestricted. The tip should be touched to the wet surface of the receiving vessel and kept in contact with it until the emptying is complete. Under no circumstance should the small amount remaining in the tip be blown out.

Measuring and serological pipettes should be held in the vertical position. After outflow has stopped, the tip should be touched to the wet surface of the receiving vessel. No drainage period is allowed. Where the small amount remaining in the tip is to be blown out and added, this will be indicated by a frosted band near the top of the pipette.

Use of a pipette filler or pipette bulb is recommended to draw the sample into a pipette. Never pipette chemical reagent solutions or unknown water samples by mouth. Use the following techniques for best results:

1. Draw liquid up into the pipette past the calibration mark.

2. Quickly remove the bulb and place dry finger tip over the end of the pipette.

3. Wipe excess liquid from the tip of the pipette using laboratory tissue paper.

4. Lift finger and allow desired amount, or all, of liquid to drain.

NOTE: There are pipette bulbs with valves that can control the flow of liquid from the pipette without removing the bulb.

Two kinds of pipettes are generally used. Those with a graduated stem, called *Mohr pipettes*, can be used to measure any volume up to the capacity of the pipette. Those with a single measuring ring near the top are called *volumetric* or *transfer pipettes*. Typical Mohr and volumetric pipettes are shown in Figure 3–8.

Pipettes marked with the letters "TD" are designed to deliver the calibrated volume of the pipette. They will deliver the specified amount if the following conditions are met:

• The pipette is clean.

• The pipette is held in a near-vertical position during delivery.

FIGURE 3–8 Mohr pipette *(top)* and volumetric pipette *(bottom)*

- Drainage is allowed to continue 5 seconds after the level of liquid in the tip appears constant.
- Contact is made between the tip of the pipette and the receiving vessel at the end of the transfer.

The small drop of solution that will be left in the pipette is accounted for in the pipette's calibration. If a pipette has two bands ground into the glass at the top, it has been calibrated for the last drop in the pipette to be blown out.

Pipettes are constructed with the delivery end tapered and the opposite end fire-polished so that it can be closed easily with a fingertip. For work requiring great accuracy, samples should be measured with a volumetric pipette. If the sample to be measured is less than 50 mL, it is good practice to use a pipette rather than a graduated cylinder. In general, transfer or volumetric pipettes should be used when a great deal of accuracy is required. Measurement, or Mohr, pipettes may be used when less accuracy is required.

Mouth suction should never be used to pipette solutions. Instead, a rubber suction bulb should be used.

Graduation marks on pipettes must be legible and permanently bonded to the glass. Pipettes should not be allowed to stand overnight in caustic or detergent solutions because they may become cloudy or frosted. If pipettes become badly etched or the tips become chipped, they should be discarded because such damage can interfere with accurate measurement.

Porcelain Dishes

Porcelain labware has long been favored for use at elevated temperatures. Glazed porcelain is nonporous and highly resistant to heat, sudden changes in temperature, and chemical attack. Most evaporating dishes and filtering crucibles (called *Gooch crucibles*) are made of porcelain. These dishes are used for total suspended solids (TSS) analysis and total dissolved solids (TDS) analysis. A porcelain evaporating dish and a filtering crucible are shown in Figure 3–9.

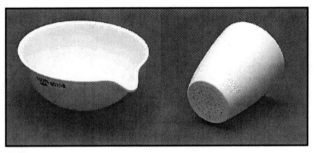

Courtesy of CoorsTek

FIGURE 3–9 Evaporating dish *(left)* and filtering crucible *(right)*

Reagent Bottles

Reagent bottles (Figure 3–10) are made of borosilicate glass because they must be stable and resistant to heat and mechanical shock. The caps, also made of borosilicate glass, are usually ground-glass stoppers with flat tops, grip tops, or penny-head tops. Tops may also be plastic.

Reagent bottles should be used exclusively for storing reagents in the laboratory. They should be clearly labeled with the following information:

- Name of the chemical and chemical formula
- Concentration of the chemical
- Date the reagent was prepared or received
- Initials of the person who prepared or received the reagent
- Expiration date of the reagent

Some reagent bottles are supplied with etched or raised-glass letters; others have a specially ground area for marking. Some reagents, such as those that are fluoride based, should be stored only in plastic reagent bottles.

Sample Bottles

Wide-mouth sample bottles are used for water sample collection, primarily because they are easier and quicker to fill than narrow-mouth bottles. In bacteriological sampling, there is less chance of contamination by splashing if wide-mouth bottles are used. Glass sample bottles should be made of borosilicate or corrosion-resistant glass, with metal or plastic closures equipped with nontoxic, leakproof liners (Figure 3–11).

FIGURE 3–10 Reagent bottle

FIGURE 3–11 Sample bottle

Plastic bottles for bacteriological and chemical samples offer the advantages of being inexpensive, lightweight, breakage resistant, and, depending on type, disposable. Autoclavable polypropylene bottles are available for collecting microbiological samples, but they should be discarded when they become brittle or discolored.

Bottles used to collect water for organic chemical analysis must be specific for the intended analysis, i.e., volatile organic compounds, synthetic organic compounds. Only borosilicate glass, polytetrafluoroethylene (PTFE; the trade name is Teflon^{TM}), or stainless-steel labware should be used. Plastic labware, such as polyethylene and polypropylene, is not acceptable for organic chemical analysis. However, plastic caps with PTFE liners may be used.

Test Tubes and Culture Tubes

Test tubes are hollow, slender glass tubes with rounded bottoms, open tops, and flared lips (Figure 3–12). Culture tubes are similar but have plain lips (Figure 3–13).

Test tubes and culture tubes can be used for a variety of general laboratory tests. They may be made of disposable plastic, disposable glass, heat-resistant glass, or special-purpose glass.

Uses of culture tubes include multiple-tube fermentation tests for bacteria, biochemical tests for bacterial identification, and stock culture collections.

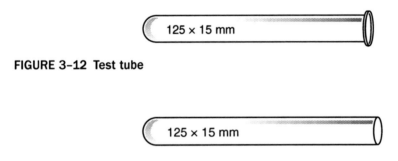

125 × 15 mm

FIGURE 3–12 Test tube

125 × 15 mm

FIGURE 3–13 Culture tube

Cleaning Labware

It is important to clean labware as soon as possible after use. This practice will ensure an adequate supply of clean labware and will promote clean labware by preventing the formation of stains. Pipettes and burettes, for example, should be rinsed promptly after use. Good labware-cleaning procedure involves two washes and two rinses:

- Detergent wash
- Acid wash with 10 percent hydrochloric acid
- Hot tap-water rinse
- Distilled-water rinse

Any good nonphosphate household detergent is adequate for cleaning most labware. Special detergents are also available from laboratory supply outlets. Liquid detergents are preferable to nonliquid types.

Dissolved matter should not be allowed to dry on labware, because if it is not completely removed, it could contaminate future analyses. If stubborn stains or crusty chemical residues remain after normal cleaning procedures, glassware should first be washed with a cleaning solution, such as an acid-dichromate type. These solutions are available either ready-made or as concentrates.

Labware used in organic chemical analysis must be cleaned with particular care—even trace amounts of organic contaminants must be removed. Specific labware should be dedicated for use only in organic chemical analysis.

MAJOR LABORATORY EQUIPMENT

For a laboratory to operate properly and perform basic analyses, a number of major pieces of equipment must be available. These are described in the following sections.

Colony Counters

A colony counter (Figure 3–14) is used to count bacterial colonies for the HPC test. Commercially manufactured colony counters magnify and backlight petri dishes so that bacterial colonies grown in the dishes can be identified and counted. Colony counters generally contain a black contrast background with a ruled counting plate to make counting easier. The viewing area is illuminated from below the culture dish. The viewing field is magnified 1.5 times by a 5-in. (130-mm) magnifying glass.

Courtesy of Reichert, Inc.

FIGURE 3–14 Colony counter

Desiccators

A desiccator is a sealable container used to hold items before they are weighed on an analytical balance. The desiccator serves two important functions: (1) it provides a place where heated items can cool slowly prior to weighing, and (2) it provides a dust- and moisture-free environment so that items being cooled will not gain moisture or contaminant weight before they are weighed. A chemical (such as dry calcium sulfate) placed in the bottom of the desiccator removes moisture from the air within.

Glass desiccators with tight-fitting glass covers and ground-glass flanged closures (Figure 3–15) are the most popular type. Desiccating cabinets made of fiberglass or stainless steel and glass are also used by large laboratories for storing opened containers of media.

Fume Hoods

A fume hood is a large enclosed cabinet that contains a fan to vent fumes out of the laboratory. When used properly, the hood is one of the most important devices for preventing laboratory accidents. A typical fume hood (Figure 3–16) contains a glass or clear acrylic (trade name Plexiglas™) door that can be closed to isolate the contents under the hood from the main laboratory. A convenient fume hood arrangement includes waste drains, electrical outlets, gas taps, and air and vacuum pressure taps, all located within the fume hood cabinet.

Courtesy of Corning Life Sciences
FIGURE 3–15 Glass desiccator

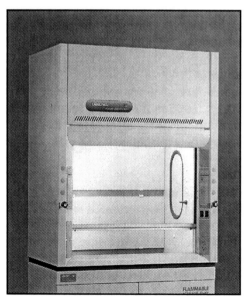

Photo reprinted with permission of Labconco Corp.
FIGURE 3–16 Fume hood

All tests that produce unpleasant or harmful smoke, gas, vapors, or fumes should be conducted under a fume hood. Whenever heat is used in a test procedure, the test should be conducted under a fume hood. The hood contains the fumes, and the hood door, if partially lowered, can protect the operator's face and upper body from accidental splashing while the test is being performed.

Incubators

An incubator is an artificially heated container used in growing bacterial cultures for microbiological tests. The three most common types of incubators are dry-heat incubators, low-temperature incubators, and water-bath incubators.

Dry-heat incubators

Dry-heat incubators contain a heating element capable of holding temperatures to within ±0.5°C of the desired incubation setting. They are useful for total coliform and HPC analyses that require a temperature of 35 ± 0.5°C. These incubators usually have a temperature range of 30 to 60°C. Because they contain a heating element only, they cannot hold temperatures below room temperature.

There are two types of dry-heat incubators: gravity convection and mechanical convection. Mechanical convection incubators (Figure 3–17) have air-circulating fans that help keep a constant temperature throughout the interior and therefore are more effective in maintaining temperature tolerance limits than are gravity-convection incubators.

Low-temperature incubators

Low-temperature incubators are used for incubation at temperature ranges from −10° to 50°C with a ± 0.3°C uniformity. These incubators are refrigerators that contain a heating element and a thermostat. They are most frequently used for biochemical oxygen demand determinations.

Water-bath incubators

Water-bath incubators are used for maintaining a more constant incubation temperature than is possible with dry-heat incubators. They are also used for many common analyses in which reactions must be completed with reagents or mixtures at a specified temperature. Water baths used as incubators for fecal coliform analyses must maintain a constant temperature of 44.5 ± 0.2°C. Water baths are capable of limiting variation from the desired setting to ± 0.2°C if the bath is covered and the water is circulated or gently agitated.

Most standard water baths are equipped only with heating elements to control temperature. These units operate in a range from room temperature to 100°C. Water baths that have both refrigeration and heating elements and an operating range of from 0 to 100°C are also available.

Courtesy of Precision/Napco

FIGURE 3–17 Mechanical convection

Jar Test Apparatus

A jar test apparatus is an automatic stirring machine equipped with three to six stirring paddles and a variable-speed motor drive. The stirring machine is mounted on top of a floc illuminator, as shown in Figure 3–18. The illuminator provides the light needed for a clear visual inspection of the floc produced during the jar test. Use of the jar test apparatus is discussed in chapter 5 and also in another book in this series, *Water Treatment*.

Membrane Filter Apparatus

A membrane filter is capable of filtering particles as small as 0.45 µm from water. A typical apparatus consists of three basic parts: a filter holder base, a membrane filter, and a filter funnel. The apparatus fits on top of a vacuum filter flask (Figure 3–19) or on a suitably designed vacuum manifold (Figure 3–20). The filter holder base is available in stainless steel, fritted glass, and plastic. The funnel is available in heat-resistant glass, plastic, or stainless steel. In addition to use in coliform analysis, the membrane filter apparatus can be used in many tests requiring preparation by filtration.

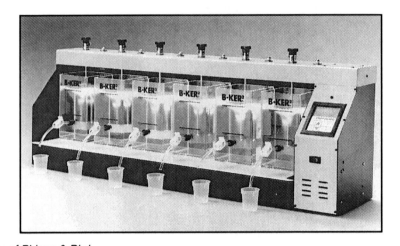

FIGURE 3–18 jar test apparatus

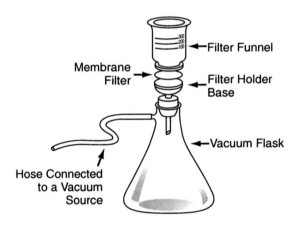

FIGURE 3–19 Membrane filter apparatus on top of a vacuum filter flask

Ovens

An oven is required in a laboratory primarily to dry, burn, or sterilize. The most commonly used ovens are utility ovens, muffle furnaces, and autoclaves.

Utility ovens

Utility ovens (Figure 3–21) typically have an operating temperature range from 30 to 350°C. They can be of two types: gravity convection or forced air. In addition,

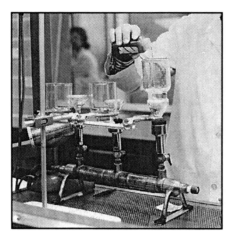

FIGURE 3-20 Membrane filter apparatus on top of a vacuum manifold

Courtesy of Barnstead International
FIGURE 3-21 Utility oven

some models are constructed so that a vacuum can be applied. These ovens are used for drying samples and labware at 105°C prior to weighing or for sterilizing labware at 170°C for use in bacteriological testing.

Muffle furnaces

Muffle furnaces (Figure 3–22) are high-temperature ovens used to ignite or burn solids. The weight of the volatile materials is found by subtracting the weight after ignition from the weight before ignition. Muffle furnaces are lined with firebrick and generally have small ignition chambers. They usually operate at temperatures near 600°C.

Autoclaves

Autoclaves are pressure cookers that are used to sterilize such items as glassware, sample bottles, membrane filter equipment, culture media, and contaminated discard materials (Figure 3–23). They sterilize by exposing the material to steam at 121°C and 15 psi (100 kPa) for a specified period of time. Exposure time varies with the kind of material to be sterilized. The use of presterilized disposable equipment may eliminate the need for an autoclave.

Courtesy of Barnstead International
FIGURE 3-22 Muffle furnace

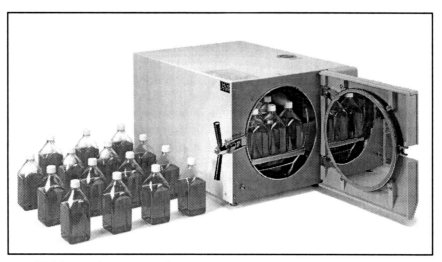

Courtesy of Brinkmann Instruments, Inc.

FIGURE 3–23 Autoclave

Refrigerators

A refrigerator is required in a laboratory to store chemical solutions and to preserve samples. A wide range of laboratory refrigerators is available, but standard domestic refrigerators are sufficient for most facilities. For bacteriological sample storage, a refrigerator capable of maintaining a temperature between 1° and 5°C is required.

Chemical solutions and samples should not be stored in the same refrigerator. Separate storage minimizes the chance of cross contamination. Food should never be kept in a refrigerator that is used for sample or chemical storage.

SAFETY EQUIPMENT

Chemical burns and fires are common laboratory hazards. Every laboratory should be equipped to protect laboratory personnel from chemical burns and to extinguish small fires.

Eye Protection

The eyes are the most vulnerable part of the human body and should be protected. Safety goggles or protective face shields should be worn when there is danger of flying particles or spattering liquids. Although prescription glasses can be purchased with shatterproof lenses, they do not surround the eyes with a tight-fitting covering to protect against splashes, nor do regular safety glasses.

Chemical splash goggles (Figure 3–24) and full-face shields (Figure 3–25) are specifically designed to reduce the chance of liquids reaching the eye. Also, the lens material is resistant to impact and penetration. Both types of eye protectors can be worn over normal prescription glasses.

Contact lenses can increase eye injury from chemical splashes, and it is recommended that they not be worn in laboratories.

An eyewash should be available in every laboratory. When a highly alkaline or acidic chemical touches the eyes or skin, deterioration begins immediately; the longer the period of contact, the greater the damage that will occur. The eyewash quickly floods the eye with water. Eyewashes can consist of bottles with an eye cup or spray nozzle designed to flood the eye (Figure 3–26), or they can be permanent plumbing fixtures similar to drinking fountains (Figure 3–27).

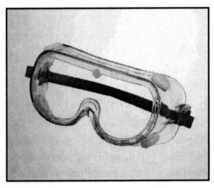

Courtesy of Bel-Art Products, Inc.

FIGURE 3–24 Safety goggles

Courtesy of Thomas Scientific, Inc.

FIGURE 3–25 Full-face shield

Courtesy of Bel-Art Products, Inc.

FIGURE 3–26 Eyewash bottle

FIGURE 3–27 Eyewash apparatus similar to a drinking fountain

Deluge Showers

Deluge showers deliver a torrent of water in a uniform pattern to wash a person's body as completely and as rapidly as possible. As shown in Figure 3–28, a free-standing deluge shower can be placed in a convenient, easy-to-reach location in the laboratory. The shower should have a large, easy-to-grab pull-chain ring or a paddle valve. Once the shower is turned on, it should remain on until deliberately turned off.

Fire Extinguishers

Quick use of a fire extinguisher can prevent a small laboratory fire from becoming a large one. Every laboratory should have at least one all-purpose fire extinguisher (Figure 3–29) capable of putting out small fires. Laboratories should also be equipped with a fire blanket. The blanket's major purpose is to extinguish burning clothing, but it can also be used to smother liquid fires in small open containers. The blanket (Figure 3–30) is usually stored in a container mounted on a wall or column and is arranged in the container so that it can easily be pulled out. The fire extinguisher and fire blanket can extinguish almost all small fires that commonly occur in the laboratory.

FIGURE 3–28 Deluge shower

FIGURE 3–29 Fire extinguisher   FIGURE 3–30 Fire blanket

Water Stills and Deionizers

Two types of high-purity water are commonly used in most laboratories:

1. Distilled water
2. Deionized water

Water stills

A water distillation unit (still) produces the distilled water needed for many laboratory tests and for rinsing all labware prior to use. Stills like the one shown in Figure 3–31 produce distilled water from common tap water by evaporation and condensation. Distilled water is free of dissolved minerals, uncombined gases, and all types of organic and inorganic nonvolatile contaminants. Stills can be portable or fixed and are generally made of glass for laboratory use. They can be heated by gas, electricity, or steam. Laboratory-size stills generally have an output capacity from about 0.3 to 5 gph (1.1 to 18.9 L/hr).

Deionizers

A deionizer removes all dissolved inorganic material (ions) from water using special ion exchange resins. Organic matter, chlorine, uncombined gases, and fine particulates are not removed unless the unit is equipped with special cartridges for their removal. As shown in Figure 3–32, deionizers are available in simple cartridge form for direct connection to any laboratory water faucet. The units continue to produce deionized water until the resin becomes exhausted. Exhaustion of the deionizer is signaled by a change in resin color or by a light or meter provided with the unit. When exhausted, the old cartridge is removed and a new one inserted in its place. Deionizers have capacities up to 60 gph (227 L/hr).

Courtesy of Barnstead International
FIGURE 3-31 Water still

Courtesy of Barnstead International
FIGURE 3-32 Deionizer

Deionized water can be used for most general laboratory purposes, including the preparation of solutions and washing of precipitates, extraction, and rinsing of glassware in place of distilled water. Deionized water cannot be substituted for distilled water where organic impurities will interfere with an analytical method unless the unit is also equipped with an organic chemical removal cartridge.

SUPPORT EQUIPMENT

A well-equipped laboratory includes a variety of support equipment used for various tests.

Aspirators

An aspirator is a T-shaped plumbing fixture that connects to a water faucet and is used to create a vacuum. A typical glass aspirator is shown in Figure 3–33. Metal or plastic units are best for use in a sink. When the faucet is turned on, water flows down the vertical leg of the aspirator, creating a negative pressure (vacuum) in the horizontal stem. When connected to a vacuum filter flask, an aspirator produces the vacuum needed for many laboratory filtering operations.

Aspirators can create a cross-connection and a potential hazard to the laboratory water supply. The faucet used to connect an aspirator should be provided with an atmospheric vacuum breaker.

FIGURE 3–33 Aspirator

Hot Plates

Hot plates are widely used in laboratories for heating solutions. Hot plates usually have a temperature range of 100° to 500°C. The heating surface is smooth for easy cleaning and is made of a corrosion-resistant material, such as glass, ceramic, or aluminum. A hot plate with a built-in stirring unit is illustrated in Figure 3–34.

Burners

A gas (Bunsen) burner is a convenient high-temperature heating device used in any laboratory served by natural gas or equipped with bottled gas (Figure 3–35). Burners are provided with adjustable air intake shutters for proper air-and-gas mixing.

Filters

Three types of filters commonly used in laboratories today are filter paper, glass-fiber filters, and membrane filters.

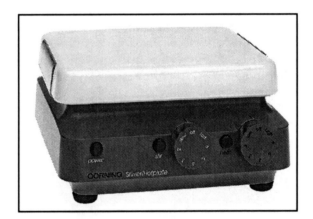

Courtesy of Corning Life Sciences

FIGURE 3–34 Hot plate with built-in stirring unit

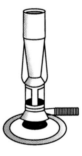

FIGURE 3–35 Gas burner

Filter paper

Filters made of porous paper are used to clarify solutions, collect particulates, and separate solids from liquids. The paper's pore size is between 5 and 10 µm, and the filters are available in diameters ranging from 4.25 to 50 cm.

Glass-fiber filters

Filters made of uniform fine glass fibers are used to filter fine particulates, bacteria, and algae under a high rate of flow. Pore size varies from 0.7 to 2.7 µm, and diameters from 15 to 261 cm are available.

Membrane filters

Filters made of cellulose acetate create membranes with precise pore sizes ranging from 0.2 to 5.0 µm. They are available in diameters from 13 to 142 mm. They have many uses in the laboratory; however, the main use is for bacterial testing.

Magnetic Stirrers

Magnetic stirrers are used to stir solutions continuously for long periods of time. They are similar in appearance to laboratory hot plates, having a corrosion-resistant top of aluminum, glass, or ceramic material. Beneath the top is a variable-speed motor that drives a rotating magnetic field. This magnetic field spins a magnetized, plastic-coated stirring bar that is placed in the liquid to be stirred. Combination magnetic stirrer–hot plate units contain separately controlled heating elements and stirring mechanisms. The units can function as heaters, stirrers, or both.

Vacuum Pumps

Vacuum pumps are commonly used in laboratories to aid in filtration. A large laboratory may have a large vacuum pump connected by pipes to taps in different areas of the laboratory. Smaller laboratories find small, portable pressure-and-vacuum pumps suitable (Figure 3–36). These portable units typically use a ⅛-hp to ⅓-hp electric motor and can produce vacuums of as much as 28 in. (94 kPa) of mercury and pressures of as much as 50 psig (350 kPa, gauge). The attached filters, shown in Figure 3–36, help to provide oil-free output air.

ANALYTICAL LABORATORY INSTRUMENTS

Many sensitive instruments are required in a water laboratory to measure various parameters. Some of the more common ones are described in the following paragraphs.

Courtesy of Gast Manufacturing, Inc.

FIGURE 3–36 Pressure-and-vacuum pump

Balances

The balance is a precise instrument used to measure weight. The pan balance, or "rough" balance, weighs loads as much as approximately 2 kg. Pan balances are available in single- and double-pan models.

Single-pan balances

To use the single-pan balance, the item to be weighed is placed on the pan and the counterweights, located on the three horizontal arms (beams), are adjusted. The indicator arrow on the far right end of the three beams will show when the counterweights equal the weight of the item. The weight reading is obtained by adding the amount of weight on each of the three beams.

Double-pan balances

The double-pan balance has the same capability as the single-pan balance, but the procedure is more time consuming. Standard brass weights must be placed on the right-hand pan until the two pans can be balanced by sliding a weight across a beam. The weight of the item is found by adding to the beam reading the total of standard brass weights used.

Analytical balances

Analytical balances are far more sensitive and precise than pan balances and can detect a change in weight of as little as ±0.0001 g (0.1 mg). The most convenient and practical analytical balance used in today's laboratory is the automatic single-pan balance (Figure 3–37). The word *automatic* refers to the built-in standard weights, which are placed in operation quickly and "automatically" by simply turning a knob. The final weight is easily read from a display. This system simplifies the weighing procedure and minimizes errors in recording weights. The rapidity of the weighing is an important feature of the analytical balance because items being weighed can pick up moisture from the air, causing a slow weight gain.

Digital balances

Top-loading digital balances (Figure 3–38) are now replacing the older types of balances because of their improved accuracy and speed of operation.

Locations for balances

A balance must be located on a solid, level surface for it to function properly. Metal counters, for example, are not suitable because of the flexibility and potential movement of the metal. Vibrations from machinery operating nearby or from an unstable floor can be transmitted through the table to the balance, causing an inaccurate reading. The vibrations from machinery can be greatly reduced if a properly constructed table is located on an unyielding floor.

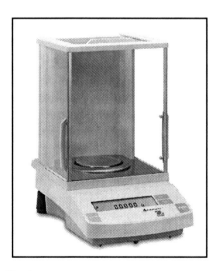

Courtesy of Thomas Scientific, Inc.

FIGURE 3–37 Automatic single-pan analytical balance

Courtesy of Ohaus Corporation (Scout® Pro)

FIGURE 3–38 Top-loading portable electronic balance

Most laboratories use solid-marble balance tables like the one shown in Figure 3–39. These tables reduce the transmission of vibrations from the floor and remain level. They should be placed on concrete floors if possible, near a bearing wall if above ground level, and in an environment with relatively constant temperature and humidity and with no sunlight.

Meters

A variety of specialized meters are used in water treatment plant laboratories for measuring water quality characteristics.

Colorimeters

The concentrations of many chemicals can be determined by measuring the intensity of color in a chemical reaction. Colorimetric measurements may be made using a wide variety of equipment, including standard color-comparison tubes, photoelectric colorimeters, and spectral photometers. Each has its place and its particular application in the laboratory.

Color comparators with permanent color standards for specific parameters can be purchased for laboratory and field use. There are two types of comparators. The disk type consists of a wheel of small colored glasses. The slide type consists of liquid standards and glass ampules. Comparators give rapid, fairly acceptable, consistent results. The most common comparators are the chlorine residual test kit, used by most water utilities, and the chlorine–pH test kit, used for swimming pools.

Courtesy of Thomas Scientific, Inc.
FIGURE 3–39 Marble balance table

Electrical conductivity meters

A common way of obtaining a quick estimate of the concentration of dissolved solids in water is to measure the electrical conductivity (EC) of the water. The EC meter actually measures the electrical resistance of the water between two electrodes suspended in the sample. The instrument readout is in microhms per centimeter at 25°C. In general, every 10 units of EC represents 6 to 7 mg/L of dissolved solids.

Photometers

The photometer is an electronic device that performs the same function as a colorimeter or color comparator but is far more accurate and precise than visual colorimeters. A photometer can measure small differences in color intensities not easily seen by the human eye. Other advantages over visual colorimeters include freedom from variable light conditions and elimination of errors because of color blindness or color bias of the analyst.

The photometer is versatile, easy to use, and relatively inexpensive. US Environmental Protection Agency (USEPA) drinking water regulations allow the use of photometers in testing for nitrate, arsenic, fluoride, and chlorine residual.

A basic photometer, such as the one in Figure 3–40, has five main components:

1. White light source
2. Wavelength control unit
3. Sample compartment
4. Detector
5. Meter

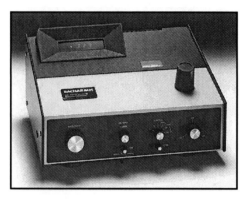

Courtesy of Bacharach, Inc., Photo by Dick Brehl

FIGURE 3–40 Photometer

The white light passes through the wavelength control unit (a simple colored filter, a diffraction grating, or a prism) to produce a single-color light (light of a specific wavelength). The single-color light then passes through the treated sample, which is contained in a glass tube (called a *cuvette*) in the sample compartment. The amount of light that passes through the sample is sensed by the detector and indicated on the graduated scale or digital meter. The measurement can be expressed either in terms of percent transmittance or in terms of absorbance. Finally, the concentration of the measured constituent is found from a previously prepared calibration curve (Figure 3–41). Such a curve must be prepared for each constituent to be measured. At regular intervals, or any time results appear suspect, a complete new set of standards should be prepared to check the calibration curve.

There are two basic types of photometers, the electrophotometer and the spectrophotometer. The basic difference between the two is the method used to produce the single-color light.

Electrophotometers. An electrophotometer uses a simple colored-glass filter. A specific filter color is required for each constituent measured. Electrophotometers are generally used for just a few difficult constituent determinations.

Spectrophotometers. A spectrophotometer uses either a diffraction grating or a prism to control the light color. When the angle of the grating or the prism is adjusted, different light colors (different wavelengths of light) can be selected. Thus, one adjustable grating or prism provides a continuous spectrum of color selections. A spectrophotometer is particularly useful when a wide variety of constituents are being measured. Its versatility allows convenient selection of the best light color for any test.

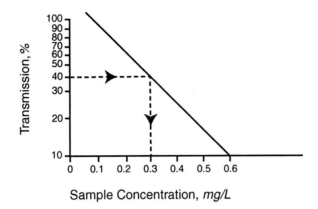

FIGURE 3–41 Calibration curve

A special type of spectrophotometer, the atomic absorption spectrophotometer (AA unit), is used for analyses of most heavy metals in water. It is a sophisticated and expensive analytical tool that must be operated by specially trained laboratory technicians.

In the AA flame procedure, the sample is aspirated into a flame, where it is vaporized. A special light source, from a specific type of hollow-cathode lamp or an electrodeless discharge lamp, emits a light at a wavelength that is characteristic of the element being measured. The amount of light absorbed by the vapor is measured. This absorbance is directly proportional to the concentration of the element in the sample.

pH meters

A pH meter is a sensitive voltmeter that measures the pH (acidity or basicity) of samples. Meters having graduated scales indicate pH units from 0 to 14. More sophisticated meters have expanded scales that allow more precise pH measurement within a narrower range and a millivolt scale that allows measurement of specific ions, such as fluoride. Many instrument types are available; a typical meter with digital readout is shown in Figure 3–42.

One or two electrodes are supplied with the meter. One electrode, a standard calomel reference type, develops a constant voltage to be compared against the changing voltage of the second. The voltage of the second electrode, a glass type, changes as the pH changes. The second electrode is designed so that a change of one pH unit produces a voltage change of 59.1 mV at 25°C. In some units, the two electrodes are mounted in a single unit called a *combined*, or *combination*, electrode.

Courtesy of Thermo Electron Corp.

FIGURE 3–42 Typical pH meter

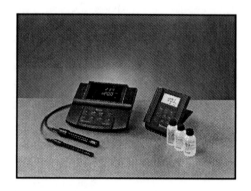

Courtesy of Thermo Eletron Corp.

FIGURE 3–43 Specific ion meter

Specific ion meters

The concentration of a specific constituent in water, such as fluoride, can be measured with a specific ion meter. The complete unit consists of a millivolt meter and interchangeable electrodes. Each electrode is selectively sensitive to one particular constituent of the water, and each specific ion test requires a different electrode. There are currently more than 20 selective electrodes, including electrodes that will measure chloride, copper, hardness, fluoride, sodium, and chlorine.

In general, most specific ion electrodes are only useful for applications in which many consecutive tests must be made on similar samples. Frequent calibrations may be necessary, often rendering the testing more time consuming than testing by other methods. Also, the electrodes are subject to interferences. The fluoride electrode is an exception; the results it obtains are excellent.

The specific ion meter (Figure 3–43) resembles a pH meter, with two major differences: the addition of a millivolt scale on the meter face and the provision for use of selective ion electrodes. Often a pH meter is purchased with a millivolt scale so that it can also be used as a specific ion meter. A specific ion meter may read concentration directly or it may read in millivolts, in which case concentration is determined by using a standard curve. When a meter with millivolt readings is being used, a standard curve to convert from millivolts to concentration must be developed. This conversion is made by measuring several samples of known concentration and plotting the results.

Turbidimeters

A turbidimeter measures the clarity of water. Basically, the amount of light impeded by or scattered by the suspended particles in the sample of water is measured. USEPA drinking water regulations approve only the nephelometric method for measuring turbidity. Therefore, although there are other methods, the nephelometric method will be the only one discussed.

Nephelometric turbidimeters are very similar to photometers, both in appearance and in performance. The turbidimeter consists of the following major components:

- Light source
- Focusing device
- Sample compartment
- Detector (photomultiplier tube)
- Meter

As shown by Figure 3–44, light passes through a focusing device into the sample compartment and through the sample. The light is reflected by the individual particles that cause turbidity. Some of that reflected light strikes a detector, such as a phototube, located 90 degrees off the main light path. The detector measures the amount of light reaching it. The meter indicates the corresponding turbidity in nephelometric turbidity units.

Nephelometric turbidimeters are quick and relatively easy to standardize and operate. Most meters have readouts that indicate turbidity values directly. The meters usually have several scale ranges; the most common ones are 0 to 0.2, 0 to 1, 0 to 10, 0 to 100, and 0 to 1,000 ntu.

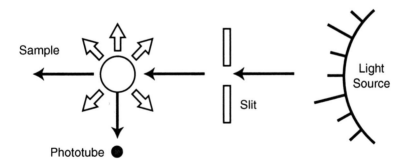

FIGURE 3–44 Path of light through a nephelometric turbidimeter

Because all communities using surface water sources are required to test their treated water daily for turbidity, the turbidimeter is a necessity at every surface water plant and is one of the most commonly used instruments.

A variation of the basic turbidimeter is the continuous monitoring online type. Instead of a sample compartment, this type has a flow-through chamber in which turbidity is continuously measured (Figure 3–45). A complete online installation typically consists of the flow-through nephelometric sensor, a meter-type turbidity indicator, and a chart recorder. Such installations are used to monitor the turbidities of raw water, in-process water, and finished water.

Microscopes

A microscope magnifies extremely small objects so that they can be seen and studied. The naked eye can see objects as small as 40 μm across, about half the diameter of a human hair. Through very powerful magnification, the microscope extends human vision into the incredibly small worlds of algae and bacteria.

Courtesy of Hach Company

FIGURE 3–45 Nephelometric turbidimeter with flow-through chamber

The simplest optical microscope is a magnifying glass. The best magnifying glass can magnify an object 20 times. This measurement is abbreviated 20×, meaning the diameter of the image is 20 times greater than the diameter of the object. The most common microscope is the compound microscope, which uses two or more lenses. A compound microscope consists of five basic parts (Figure 3–46):

1. Stand.
2. Movable stage.
3. Head, including oculars, or eyepiece lenses (a one-ocular head is called a monocular; a two-ocular head, a binocular; a triocular has one eyepiece designed for mounting a camera).
4. Objective nosepiece, a revolving set of lenses (the selection of different objectives gives different magnifications).
5. Illuminator (light source) and condenser lens used to focus light onto the object being viewed.

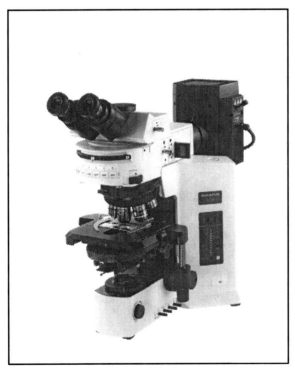

Courtesy of Olympus America, Inc.

FIGURE 3–46 Compound microscope

Typically, the compound microscope magnifies as much as 1,000×, and some advanced types reach much higher magnification. As shown in Table 3–1, the magnifying power of a compound microscope depends on the combined magnification of the eyepiece (ocular) and the objective lenses. For example, a 20× objective lens combined with a 10× eyepiece produces a magnification of 200× (20 × 10 = 200).

The compound microscope is one of the most important tools in the water quality laboratory. It is used for counting and identifying the microscopic plant and animal life typically found in water, including color-, taste-, and odor-causing algae and disease-causing bacteria.

Measurement of Organic Chemicals

Analytical instruments that measure trace levels of organic contaminants are relatively sophisticated and somewhat expensive. Many large water utility laboratories use simple gas chromatographs, but very few have gas chromatography–mass spectrophotometry (GC–MS) capability.

Gas chromatographs are essentially sophisticated distillation units; they consist mainly of ovens, columns, and detectors. The organic compounds in the sample are vaporized, moved through the columns by an inert gas, separated, and then moved to a detector. The results are displayed on a chart called a chromatograph.

A mass spectrophotometer is nothing more than a special detector on the end of a gas chromatograph. However, it has the ability to identify many organic compounds by their mass (weight). A computer is usually an integral part of a gas chromatograph or GC–MS unit. It can help identify compounds through its library of information on compound characteristics.

TABLE 3–1 Optical microscope magnification

	Overall Magnification	
Type of Objective	10× Ocular	15× Ocular
16 mm (10× or low power)	100×	150×
8 mm (20× or medium power)	200×	300×
4 mm (43× or high power)	430×	645×
1.8 mm (90× or oil immersion)	900×	1,350×

SELECTED SUPPLEMENTARY READINGS

Clark, T.F. 1988. New Culture Medium Detects Stressed Coliforms. *Opflow,* 14(3):3.

Dental, S.K., and K.M. Kingery. 1988. *An Evaluation of Streaming Current Detectors.* Denver, Colo.: American Water Works Association Research Foundation and American Water Works Association.

————. 1989. Using Streaming Current Detectors in Water Treatment. *Jour. AWWA,* 81(3):85.

Hargesheimer, E.E., and C.M. Lewis. 1995. *A Practical Guide to On-Line Particle Counting.* Denver, Colo.: American Water Works Association and American Water Works Association Research Foundation.

Hargesheimer, E.E., C.M. Lewis, and C.M. Yentsch. 1992. *Evaluation of Particle Counting as a Measure of Treatment Plant Performance.* Denver, Colo.: American Water Works Association Research Foundation and American Water Works Association.

Manual M12, Simplified Procedures for Water Examination. 2001. Denver, Colo.: American Water Works Association.

Manual of Instruction for Water Treatment Plant Operators. 1975. Albany, N.Y.: New York State Department of Health.

Manual of Water Utility Operations. 8th ed. 1988. Austin, Texas: Texas Water Utilities Association.

Standard Methods for the Examination of Water and Wastewater. 20th ed. 1998. Washington, D.C.: American Public Health Association, American Water Works Association, and Water Environment Federation.

4

Microbiological Contaminants

HISTORY

More than 2,500 years ago, Hippocrates, who is called "the father of medicine," theorized that many diseases were caused by drinking water, but he was unable to explain why. Over the ages, great epidemics caused by contaminated drinking water periodically killed large segments of populations. It was not until the nineteenth century that the germ theory was developed by researchers, such as Friedrich Henle, Robert Koch, and Louis Pasteur.

Practically all pathogenic organisms that can be carried by water originate from the intestinal tracts of warm-blooded animals, particularly humans. Some waterborne diseases can be spread by "carriers"—individuals in whose bodies the disease is active but who show few or no symptoms. One famous carrier was Mary Mallon, a woman who became known as Typhoid Mary. In the 1930s, she infected perhaps as many as 1,000 people in the United States with typhoid fever but never showed severe symptoms of the deadly disease herself.

The disease-causing organisms that are considered the principal sources of potential waterborne diseases are listed in Table 4–1. However, most of these diseases can also be transmitted by other means, such as through food or body contact. Many of the diseases that caused tremendous loss of life just 100 years ago have now been virtually eradicated in most areas of the world through a combination of improved sanitation and the use of new medications.

One famous carrier was Mary Mallon, a woman who became known as Typhoid Mary.

TABLE 4–1 Waterborne diseases

Waterborne Disease	Causative Organism	Source of Organism in Water	Symptom/Outcome
Gastroenteritis	*Salmonella* (bacteria)	Animal or human feces	Acute diarrhea and vomiting
Typhoid	*Salmonella typhosa* (bacteria)	Human feces	Inflamed intestine, enlarged spleen, high temperature—fatal
Dysentery	*Shigella*	Human feces	Diarrhea—rarely fatal
Cholera	*Vibrio comma* (bacteria)	Human feces	Vomiting, severe diarrhea, rapid dehydration, mineral loss—high mortality
Infectious hepatitis	Virus	Human feces, shellfish grown in polluted waters	Yellowed skin, enlarged liver, abdominal pain; lasts as long as 4 months—low mortality
Amoebic dysentery	*Entamoeba histolytica* (protozoa)	Human feces	Mild diarrhea, chronic dysentery
Giardiasis	*Giardia lamblia* (protozoa)	Wild animal feces suspected	Diarrhea, cramps, nausea, general weakness; lasts 1 week to 30 weeks—not fatal
Cryptosporidiosis	*Cryptosporidium*	Human and animal feces	Diarrhea, abdominal pain, vomiting, low-grade fever

In 1990, US Environmental Protection Agency's (USEPA's) Science Advisory Board (SAB), an independent panel of experts established by Congress, cited drinking water contamination as one of the most important environmental risks and indicated that disease-causing microbial contaminants (i.e., bacteria, protozoa, and viruses) are probably the greatest remaining health risk-management challenge for drinking water suppliers (USEPA/SAB 1990). Information on the number of waterborne disease outbreaks from the US Centers for Disease Control and Prevention (CDC) underscores this concern. CDC indicates that, between 1980 and 1996, 401 waterborne disease outbreaks were reported, with over 750,000 associated cases of disease. During this period, a number of agents were implicated as the cause, including protozoa, viruses, and bacteria.

Protozoa, bacteria, and viruses are microorganisms. Microorganisms are organisms too small to be seen by the naked eye and can only be seen with a microscope.

The pathogens that are still of some concern as sources of waterborne disease are discussed in the following sections.

Bacteria

Bacteria are single-cell microscopic organisms that have a definitive cell wall and a single strand of deoxyribonucleic acid (DNA). They multiply by binary fission. Bacteria are not photosynthetic like plants and do not need sunlight to multiply. They have no well-defined nucleus and do not contain chlorophyll. The more common shapes are spheres, rod-shaped, spiral and branching threads, or filamentous. Bacteria range in size from 0.5 to 1 µm in diameter to 2 to 4 µm in length. Most have flagella, a tail-like structure for movement.

Pathogenic bacteria of interest in drinking water are *Salmonella, Escherichia coli* or *E. coli, Shigella, Legionella,* and opportunistic pathogens.

Salmonella paratyphi causes typhoid. Enteropathogenic *E. coli* causes gastroenteritis in humans, most notably diarrhea. *Shigella* causes bacillary dysentery. *Legionella* causes pneumonia-like symptoms. It is often found in cooling towers and colonizes plumbing systems. *Legionella's* route to infection is through aerosol from showerheads.

Opportunistic pathogens are not normally a danger to persons in good health, but they can cause sickness or death in those who are in a weakened condition. Particularly at risk are newborns, the elderly, and persons who already have a serious disease.

Included among the opportunistic bacteria are *Pseudomonas, Aeromonas hydrophila, Edwardsiella tarda, Flavobacterium, Klebsiella, Enterobacter, Serratia, Proteus, Providencia, Citrobacter,* and *Acinetobacter.* These organisms are prevalent in the environment, and it is presently considered impossible to eliminate them completely from treated water.

Viruses

Viruses are extremely complex molecules that typically contain a protein coat surrounding a DNA or ribonucleic acid (RNA) core of genetic material. Viruses have no independent metabolism and depend on living cells for reproduction. They range in size from 10 to 25 nanometers (nm) in diameter, which is smaller than can be seen with an optical microscope.

Viruses do not live long outside of a human's or an animal's body, but while they are alive they can survive heat, drying, and chemical agents. Viruses are much more resistant to chlorine in water than are bacteria.

Some types of viruses have caused acute epidemics of gastroenteritis. The waterborne hepatitis A virus (HAV) is the source of some of the most serious health problems. HAV causes infectious hepatitis, which can result in serious liver damage or death. The CDC documented 23 outbreaks of disease caused by HAV between 1971 and 1985.

Protozoa

Protozoa, amoebas, flagellated protozoans, ciliates, and sporazoans are single-cell microorganisms with a true nucleus but without a cell wall. They reproduce by fission and feed on bacteria. Protozoa range in size from 5 to 100 μm. *Giardia lamblia* and *Cryptosporidium* are the protozoa of major concern in drinking water.

Giardia lamblia

Giardiasis is the most frequently diagnosed waterborne disease in the United States. Symptoms include skin rash, flu-like problems, diarrhea, fatigue, and severe cramps. The symptoms may last anywhere from a few days to months. Sometimes there are periods of remission when there are no symptoms, and then the illness recurs. The protozoan attaches itself to the upper intestinal tract and produces cysts, which are shed in the feces.

One of the major reasons why giardiasis continues to be a problem as a waterborne disease is that the cysts survive well under adverse conditions. They are highly resistant to chlorine and can live in cold water for months. Three of the major hosts for *Giardia* are humans, beaver, and muskrat. Although water can be a major means of transmitting the disease, the largest percentage of recorded cases is caused by person-to-person contact.

Cryptosporidium

Cryptosporidium is a parasite that has caused several outbreaks of cryptosporidiosis and poses serious health risks. Ten species are currently recognized. *Cryptosporidium parvum* is the most common form found in humans.

Waterborne diseases are usually described as acute, which means that the symptoms are sudden but in healthy people only last a short time. Most waterborne pathogens cause gastrointestinal illness with diarrhea, abdominal discomfort, nausea, vomiting, and other symptoms. Some waterborne pathogens can be associated with more serious disorders, such as hepatitis, gastric cancer, peptic ulcers, myocarditis, swollen lymph glands, meningitis, encephalitis, and many other diseases. In healthy individuals, cryptosporidiosis is an infection that usually causes 7–14 days of diarrhea with possibly a low-grade fever, nausea, and abdominal cramps. The effects on immunocompromised individuals can be life threatening. No antibiotic treatment currently exists for cryptosporidiosis.

Oocysts may be found in all types of water, including untreated surface water and filtered swimming-pool water. Outbreaks can be caused by contamination of food and of water in swimming pools and sprinklers. The 1993 outbreak in Milwaukee resulted in the death of several immunocompromised patients and illness in many previously healthy people.

Prevention
Cryptosporidium infections are contracted by the ingestion of oocysts, and therefore effective control measures must aim to reduce or prevent oocyst transmission. *Cryptosporidium* oocysts are resistant to most disinfectants and antiseptics. Aldehyde, ammonia, alcohol, chlorine, and alkaline-based commercial disinfectants are ineffective. Most conventional water treatment methods do not effectively remove or kill all oocysts. Severely immunocompromised persons are encouraged to avoid any contact with water in lakes and streams and should not drink such water. These individuals are encouraged to boil all water before consumption.

INDICATOR ORGANISMS
The tests required to detect specific pathogens are still considered time intensive and expensive, so it is impractical for water systems to routinely test for specific types of organisms. A more practical approach is to examine the water for indicator organisms specifically associated with contamination. An indicator organism essentially provides evidence of fecal contamination from humans or warm-blooded animals. The criteria for an ideal indicator organism are that it should

- always be present in contaminated water;
- always be absent when fecal contamination is not present;
- generally survive longer in water than pathogens;
- be easy to identify.

The coliform group of bacteria meets all criteria for an ideal indicator. These bacteria are generally not pathogenic, yet they are usually present when pathogens are present. Additionally, coliform bacteria are more plentiful than pathogens and can often stay alive in the environment for longer periods of time.

Coliform bacteria are easily detected in the laboratory. As a rule, where coliforms are found in water, it is assumed that pathogens may also be present, making the water bacteriologically unsafe to drink. If coliform bacteria are absent, the water is assumed safe.

Four methods exist for determining the presence of coliform bacteria in a water sample: the multiple-tube fermentation (MTF) method, the presence–absence (P–A) method, the MMO–MUG method, and the membrane filter (MF) method. Detailed descriptions of the analytical procedures for these tests can be found in the latest edition of *Standard Methods for the Examination of Water and Wastewater*.

113

Coliform Analyses

The detection of coliform bacteria in a water sample by any of the four analytical techniques is a warning of possible contamination. One positive test does not conclusively prove contamination, however, and additional tests must be conducted. Samples are often contaminated by improper sampling technique, improperly sterilized bottles, and, on rare occasions, laboratory error. Regulatory agencies recognize this fact, and drinking water regulations require check or repeat sampling after findings that show a positive test for coliform in a sample. Drinking water regulations and maximum contaminant levels for coliform bacteria are discussed in chapter 1 of this book.

Sampling

Sterile containers must be used for all samples collected for bacteriological analysis. The same sampling procedures should be used for coliform analysis and heterotrophic plate count (HPC) analysis (refer to chapter 2 of this book). (See Table 4–2.)

Test Methods

The principle of operation of the MTF and P–A tests utilizes the fact that coliform bacteria produce a gas within 48 hours when incubated at 35°C. Although the bacteria themselves cannot be seen, the gas that is formed and trapped in an inverted vial in the fermentation tube signifies their presence.

Multiple-tube fermentation method

The MTF test (Figure 4–1) progresses through three distinct steps:

1. Presumptive test
2. Confirmed test
3. Completed test

Presumptive test. The presumptive test is the first step of the analysis. Samples are poured into each of five tubes containing a culture medium (lactose or lauryl tryptose broth [LTB]) and the vials are inverted. The samples are then incubated for 24 hours, checked, incubated for 24 hours more, and checked again.

If coliform bacteria are present in the water, the gas they produce will begin to form a bubble in each inverted vial within the 48-hour period; this is called a *positive sample* or reported as *presence*. If no gas forms, the sample is called *negative* or reported as *absence*. If gas is produced after either the 24-hour or the 48-hour incubation period, the sample must undergo the confirmed test.

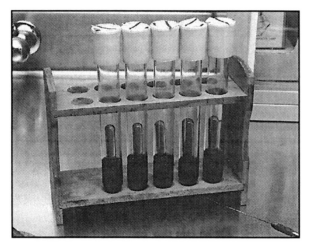

Source: Opflow *(May 1983)*

FIGURE 4-1 Typical multiple-tube fermentation setup

Confirmed test. The confirmed test is more selective for coliform bacteria. This test increases the likelihood that positive results obtained in the presumptive test are caused by coliform bacteria and not other kinds of bacteria. Cultures from the positive samples in the presumptive test are transferred to brilliant green lactose bile (BGB) broth and incubated. If no gas has been produced after 48 hours of incubation, the test is negative and no coliform bacteria are present. If gas is produced, the test is positive, indicating the presence of coliform bacteria.

Bacteriological testing of most public water supplies stops after the confirmed test. This is the minimum testing that all samples must undergo when the MTF method is used. To check its procedures, the laboratory should conduct the completed test on at least 10 percent of the positive tubes from the confirmed test.

Completed test. The completed test is used to definitely establish the presence of coliform bacteria for quality control purposes. A sample from the positive confirmed test is placed on an eosin methylene blue (EMB) agar plate and incubated. A coliform colony will form on each EMB plate.

A small portion of the coliform colony is transferred to a growth medium and incubated for 18 to 24 hours. A second portion is transferred to an LTB and incubated for 24 to 48 hours. The completed test is positive if (1) gas is produced in the LTB, and (2) red-stained, nonsporeforming, rod-shaped bacteria are found. If no gas is produced in the LTB or if red-stained, chainlike cocci or blue-stained, rod-shaped bacteria are found on the agar, then the test is negative. Figure 4–2 provides a summary of the MTF method.

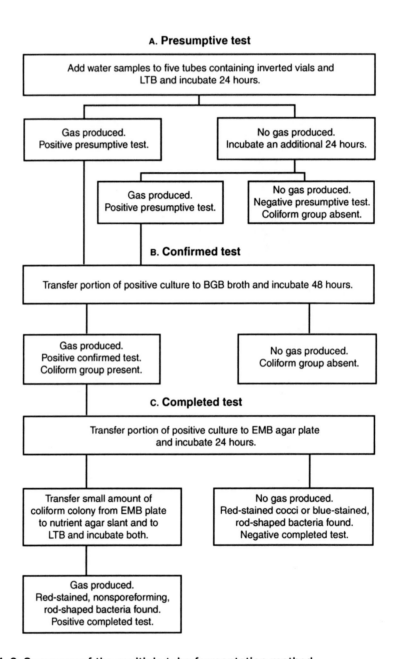

A. **Presumptive test**

Add water samples to five tubes containing inverted vials and LTB and incubate 24 hours.

Gas produced.
Positive presumptive test.

No gas produced.
Incubate an additional 24 hours.

Gas produced.
Positive presumptive test.

No gas produced.
Negative presumptive test.
Coliform group absent.

B. **Confirmed test**

Transfer portion of positive culture to BGB broth and incubate 48 hours.

Gas produced.
Positive confirmed test.
Coliform group present.

No gas produced.
Coliform group absent.

C. **Completed test**

Transfer portion of positive culture to EMB agar plate and incubate 24 hours.

Transfer small amount of coliform colony from EMB plate to nutrient agar slant and to LTB and incubate both.

No gas produced.
Red-stained cocci or blue-stained, rod-shaped bacteria found.
Negative completed test.

Gas produced.
Red-stained, nonsporeforming, rod-shaped bacteria found.
Positive completed test.

FIGURE 4–2 Summary of the multiple-tube fermentation method

Presence–absence test

The P–A test is a simple modification of the MTF method. It is intended for use on routine samples collected from a distribution system or water treatment plant. A 100-mL portion of the sample is inoculated into a 250-mL milk dilution bottle containing special P–A media and a small inverted tube. The sample is then incubated at 35°C for 24 and 48 hours.

The presence of total coliforms is indicated by the purple P–A medium turning yellow (indicating acid production) and the formation of gas in the medium. All yellow and gas-producing samples from this presumptive stage must then be confirmed as in the MTF-confirmed step, using BGB tubes. Gas production indicates the presence of total coliforms and must be reported as a positive (presence) sample in the monthly report to the primacy agency.

Samples confirmed for total coliforms must also be analyzed for either fecal coliforms or *E. coli*. A positive result from this analysis is considered an acute violation of the Total Coliform Rule (TCR) and must be reported to the primacy agency within 24 hours after results become known.

Fecal coliform procedure. If the MTF or P–A method is being used, as the presumptive positive samples are being inoculated into the BGB broth, 0.1 mL of the presumptive broth is also transferred into an electrical conductivity (EC) broth tube. If the membrane filter method is used, bacterial growth is transferred into an EC tube. This tube is then incubated for 24 hours in a water bath at 44.5°C. The presence of gas in the tube confirms the presence of fecal coliforms.

E. coli procedure. The presence of *E. coli* can be determined using the MUG test discussed in the next section. A 0.1-mL portion of the presumptive media or a swab is used to transfer a sample from a membrane filter into an EC–MUG tube. A tube that fluoresces under a long-wave ultraviolet (UV) light is confirmation for *E. coli*.

MMO–MUG technique

The MMO–MUG technique was approved by USEPA shortly after the promulgation of the TCR. MMO and MUG are acronyms for the constituents in the medium used in the tests. MMO represents minimal media with ONPG. ONPG stands for *ortho*-nitrophenyl-beta-D-galactopyranoside.

E. coli produce a specific enzyme that reacts with ONPG to give a yellow color. MUG stands for 4-methylumbelliferyl-beta-D-glucuronide. Only *E. coli* produce an enzyme that reacts with MUG. Therefore, a medium containing MMO and MUG can be used to identify both total coliforms and *E. coli* in a single-sample inoculation. Two procedures may be used. In the *ten-tube procedure,* ten tubes are purchased with the medium already in them. A 10-mL portion of sample is

TABLE 4-2 Total coliform sampling requirements, according to population served

Population Served	Minimum Number of Routine Samples per Month[*]
15 to 1,000[†]	1[‡]
1,001 to 2,500	2
2,501 to 3,300	3
3,301 to 4,100	4
4,101 to 4,900	5
4,901 to 5,800	6
5,801 to 6,700	7
6,701 to 7,600	8
7,601 to 8,500	9
8,501 to 12,900	10
12,901 to 17,200	15
17,201 to 21,500	20
21,501 to 25,000	25
25,001 to 33,000	30
33,001 to 41,000	40
41,001 to 50,000	50
50,001 to 59,000	60
59,001 to 70,000	70
70,001 to 83,000	80
83,001 to 96,000	90
96,001 to 130,000	100
130,001 to 220,000	120
220,001 to 320,000	150
320,001 to 450,000	180
450,001 to 600,000	210
600,001 to 780,000	240

Table continued next page

TABLE 4–2 Total coliform sampling requirements, according to population served *(continued)*

Population Served	Minimum Number of Routine Samples per Month[*]
780,001 to 970,000	270
970,001 to 1,230,000	300
1,230,001 to 1,520,000	330
1,520,001 to 1,850,000	360
1,850,001 to 2,270,000	390
2,270,001 to 3,020,000	420
3,020,001 to 3,960,000	450
3,960,001 or more	480

Source: Water Quality and Treatment *(1999).*

[*] In lieu of the frequency specified in this table, a noncommunity water system using groundwater and serving 1,000 persons or fewer may monitor at a lesser frequency specified by the state until a sanitary survey is conducted and the state reviews the results. Thereafter, noncommunity water systems using groundwater and serving 1,000 persons or fewer must monitor in each calendar quarter during which the system provides water to the public, unless the state determines that some other frequency is more appropriate and notifies the system (in writing). Five years after promulgation of TCR, noncommunity water systems using groundwater and serving 1,000 persons or fewer must monitor at least once per year.

† Includes public water systems that have at least 15 service connections but serve fewer than 25 persons.

‡ For a community water system serving 25 to 1,000 persons, the state may reduce this sampling frequency if a sanitary survey conducted in the last 5 years indicates that the water system is supplied solely by a protected groundwater source and is free of sanitary defects. However, in no case may the state reduce the sampling frequency to less than once per quarter.

transferred into each tube and incubated at 35°C for 24 hours. In the *P–A procedure,* the medium is purchased in vials. The medium is transferred into a bottle containing 100 mL of sample, is mixed, and is incubated as in the ten-tube procedure. If total coliforms are present in either procedure, the medium will turn yellow. If *E. coli* are present, the medium will also fluoresce blue under a UV light.

Membrane filter method

The MF method of coliform testing begins with the filtering of 100 mL of sample under a vacuum through a membrane filter. The filter is then placed in a sterile container (Figure 4–3) and incubated in contact with a selective culture medium.

119

FIGURE 4-3 Placement of membrane on pad soaked with culture medium

A coliform bacteria colony will develop at each point on the filter where a viable coliform bacterium was left during filtration. After a 24-hour incubation period, the number of colonies per 100 mL is counted (Figure 4–4).

A typical coliform colony is pink to dark red with a distinctive green metallic surface sheen. All organisms producing such colonies within 24 hours are considered members of the coliform group. For further confirmation, representative colonies are inoculated into LTB and BGB broth.

HETEROTROPHIC PLATE COUNT PROCEDURE

The HPC procedure is the only practical way to estimate the total population of bacteria in water. The test determines the total number of bacteria in a sample that will grow under specific conditions in a selected medium.

Uses of the HPC Procedure

No single food supply, incubation temperature, and moisture condition suits every type of bacteria, so a standardized procedure must be used to obtain consistent and comparable results. The procedure will therefore generally reveal fewer bacteria than are actually present.

The conditions for the procedure are selected primarily to promote growth of microorganisms that are associated with animals and humans and can be considered as pollution indicators.

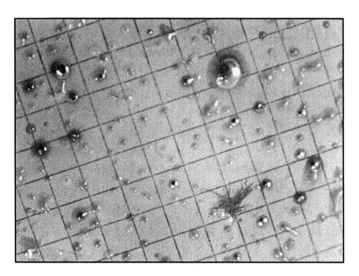

FIGURE 4–4 Membrane filter after incubation with positive growth colonies

Plate count tests are sensitive to changes in raw-water quality and are useful for judging the efficiency of various treatment processes in removing bacteria. For example, if a plate count is higher after filtration than before filtration, there may be bacterial growth on or in the filters. The problem would probably not show up during routine coliform analysis.

It is also common for water leaving a treatment plant to have a low bacterial population but for the population to have greatly increased by the time the water reaches the consumer. This occurrence is caused by bacterial aftergrowth (regrowth)—bacteria reproducing in the distribution system. Standard plate-count determinations will indicate whether this problem exists. Aftergrowth generally results from stagnant water in dead ends in the system, inadequate chlorination, or recontamination of the water after chlorination.

Performing the HPC Procedure

The HPC is performed by placing diluted water samples on plate count agar. The samples are incubated for 48 to 72 hours. Bacteria occur singly, in pairs, in chains, and in clusters. The bacteria colonies that grow on the agar are counted using bacteria-counting equipment. Detailed procedures are described in the latest edition of *Standard Methods*. These procedures must be closely followed in order to provide reliable data for water quality control measurements.

Properly treated water should have an HPC of less than 500 colonies per milliliter. Higher counts indicate an operational problem that should be investigated.

SELECTED SUPPLEMENTARY READINGS

Edberg, S.C., F. Ludwig, and D.B. Smith. 1991. *The Colilert® System for Total Coliforms and Escherichia coli.* Denver, Colo.: American Water Works Association Research Foundation and American Water Works Association.

Giardia lamblia in Water Supplies—Detection, Occurrence, and Removal. 1985. Denver, Colo.: American Water Works Association.

LeChevallier, M.W., W.D. Norton, R.G. Lee, and J.B. Rose. 1991. *Giardia and Cryptosporidium in Water Supplies.* Denver, Colo.: American Water Works Association Research Foundation and American Water Works Association.

Leland, D.E., J. McAnulty, W. Keene, and G. Stevens. 1993. A Cryptosporidiosis Outbreak in a Filtered-Water Supply. *Jour. AWWA,* 85(5):34.

Lisle, J. 1994. *An Operator's Guide to Bacteriological Testing.* Denver, Colo.: American Water Works Association.

Manual M12, Simplified Procedures for Water Examination. 2001. Denver, Colo.: American Water Works Association.

Manual of Instruction for Water Treatment Plant Operators. 1975. Albany, N.Y.: New York State Department of Health.

Manual of Water Utility Operations. 8th ed. 1988. Austin, Texas: Texas Water Utilities Association.

Miller, K.J. 1994. Protecting Customers From *Cryptosporidium. Jour. AWWA,* 86(12):8.

Moore, A.C., B.L. Herwaldt, G.F. Craun, R.L. Calderon, A.K. Highsmith, and D.D. Juranek. 1994. Waterborne Disease in the United States, 1991 and 1992. *Jour. AWWA,* 86(2):87.

Pontius, F.W. 1990. Rule Changes Way Systems Will Look at Coliform. *Opflow,* 16(12):1.

———. 1993. Protecting the Public Against *Cryptosporidium. Jour. AWWA,* 85(8):18.

Standard Methods for the Examination of Water and Wastewater. 20th ed. 1998. Washington, D.C.: American Public Health Association, American Water Works Association, and Water Environment Federation.

Testing the Microbiological Quality of Water. 1983. *Opflow,* 9(5):4–5.

Water Quality and Treatment. 5th ed. 1999. New York: McGraw-Hill and American Water Works Association (available from AWWA).

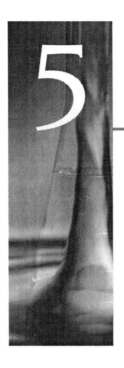

5

Physical and Aggregate Properties of Water

This chapter describes the physical properties of water. It coincides with *Standard Methods for the Examination of Water and Wastewater*, 20th edition (1998).

Physical testing of drinking water measures the physical properties of the water as distinguished from chemical or biological contaminants.

A number of physical tests are either required by regulations or necessary for control of the water treatment processes. Water systems using surface water sources perform physical tests frequently because of changing water quality. Groundwater quality generally fluctuates slowly or infrequently, so most physical testing of groundwater is performed periodically to ensure that optimum treatment is provided and that the finished water meets state and federal standards.

ACIDITY

The definition of *acidity* is base neutralizing power. There are two types of acidity: phenolphthalein acidity (CO_2 acidity), which always has a pH above 4.3 and mineral acid. Phosphoric acid, hydrochloric acid, and sulfuric acid are all mineral acids.

Methods of Measurement

Mineral acids are measured by titration to a pH of 4.3 using methyl orange as an indicator (orange on the alkaline side to salmon pink on the acid side). Mineral acidity plus acidity caused by weak acids (e.g., carbonic acid) is measured by titration of the sample to the phenolphthalein end point (pH 8.3). This is called *total acidity* or *phenolphthalein acidity*. Phenolphthalein is colorless at a pH less than 8.3 and pink or red above pH 8.3. NOTE: To a theoretical chemist, a pH of 7 is considered neutral. To a water chemist, who wants to know how much free or combined CO_2 is present and how much total alkalinity is present, a pH of 7 means very little. For a water chemist, the dividing point between acidity and alkalinity is *not* pH 7.0 but the "M" (or total) alkalinity (acidity) end point of 4.3.

Significance

Acidity plays a more important role in industrial water applications, especially in neutralization reactions. In drinking water, acids are associated with corrosive environments. Chemical reaction rates are influenced by acidity. Acids are used in the regeneration of ion exchange systems. They are also used to lower the pH of water treated by nanofiltration and reverse osmosis membranes.

ALKALINITY

The definition of *alkalinity* is acid neutralizing power (see Table 5–1).

TABLE 5–1 Alkalinity relationships

Result of Titration	Hydroxide Alkalinity	Carbonate Alkalinity	Bicarbonate Alkalinity
$P^* = 0$	0	0	MO
$P < \frac{1}{2}$ MO[†]	0	2P	MO – 2P
$P = \frac{1}{2}$ MO	0	2P	0
$P > \frac{1}{2}$ MO	2P – MO	2 (MO – P)	0
P – MO	MO	0	0

*P, phenolphthalein.
†MO, methyl orange.

Types

Phenolphthalein alkalinity ("P" alkalinity) has a pH of 8.3 and represents the end point for phenolphthalein or "P" alkalinity. Methyl orange or total alkalinity ("MO" alkalinity) has a pH of 4.3 and is directly related to the amount of hydroxide, carbonate, or bicarbonate alkalinity present.

More specifically, when only hydroxide anions are present, water samples will have a pH above 10. Titration with strong acid is complete at phenolphthalein end point (8.3). Hydroxide alkalinity is equal to phenolphthalein alkalinity.

When only carbonate is present, water samples will have a pH of 8.3 or higher. Titration to phenolphthalein end point is exactly one half of total titration to pH 4.3. Carbonate alkalinity equals total alkalinity.

When only hydroxide-carbonate is present, water samples will have a high pH, usually well above 10. Titration from pH 8.3 to 4.3 represents one half of carbonate alkalinity.

When only carbonate-bicarbonate is present, water samples will have a pH greater than 8.3 and less than 11. Titration to pH 8.3 represents one half of carbonate alkalinity.

When only bicarbonate is present, water samples will have a pH 8.3 or less (usually less). Bicarbonate alkalinity equals total alkalinity.

CALCIUM CARBONATE STABILITY

The principal scale-forming substance in water is calcium carbonate ($CaCo_3$). Water is considered stable when it will neither dissolve nor deposit calcium carbonate. This point is referred to as the *calcium carbonate stability* or *equilibrium point*. The reactions and behavior of calcium carbonate and calcium bicarbonate are therefore important in water supplies. The actual amount of calcium carbonate that will remain in solution in water depends on several characteristics of the water: calcium content, alkalinity, pH, temperature, and total dissolved solids (TDS).

Significance

Scale formation can cause serious problems in water distribution mains and household plumbing systems by restricting flow, plugging valves, and fouling hot-water heaters and boilers. Corrosion can cause premature pipe or equipment failure. Public health and aesthetic problems can also result if water is corrosive because pipe materials, such as lead, cadmium, and iron, will dissolve into the water.

Several methods can be used to determine the calcium carbonate stability of water. A popular method is the Langelier saturation index (LSI). The LSI is equal to the measured pH (of the water) minus the pH_s (saturation). The pH_s is the theoretical pH at which calcium carbonate will neither dissolve into nor precipitate from water. At the pH_s, a water is perfectly stable. Therefore, if $pH - pH_s = 0$, the water is in equilibrium and will neither dissolve calcium carbonate nor deposit it on the pipes.

If $pH - pH_s > 0$ (positive value), the water is not in equilibrium and will tend to deposit calcium carbonate on main and other piping surfaces. If $pH - pH_s < 0$ (negative value), the water is also not in equilibrium and will tend to dissolve the calcium carbonate it contacts; no coating will be deposited on the distribution pipes, and if the pipes are not protected, they may corrode.

The calcium carbonate stability of water is maintained in the distribution system by adjusting the LSI of the water to a slightly positive value so that a slight deposit of calcium carbonate will be maintained on pipe walls. Adjustment is usually made by adding lime, soda ash, or caustic soda.

Sampling

If calcium carbonate stability maintenance is used for corrosion control, finished water at the treatment plant and in the distribution system should be evaluated routinely for calcium carbonate stability. Evaluation is particularly important when treatment plant unit processes or chemical doses are changed. If the LSI indicates unfavorable conditions, process adjustments should be made. It is very important to remember that the LSI is only an indicator of stability; it is not an exact measure of corrosivity or of calcium carbonate deposition. The LSI is developed from results of alkalinity, pH, temperature, calcium content, and TDS (dissolved residue) monitoring.

Methods of Determination

If the temperature, TDS, calcium content, and alkalinity of the water are known, the pH_s can be calculated. The following equation may be used:

$$pH_s = A + B - \log (Ca^{+2}) - \log (\text{alkalinity})$$

In the equation, A and B are constants, and calcium and alkalinity values are expressed in terms of milligrams per liter as calcium carbonate equivalents. Tables 5–2, 5-3, and 5-4 are used to determine the values of the constants and logarithms.

The actual pH of the water is measured directly with a pH meter, and the LSI is calculated using the formula $LSI = pH - pH_s$.

Example:

A sample of water has the following characteristics:

Ca^{+2}　　= 300 mg/L as $CaCO_3$

Alkalinity = 200 mg/L as $CaCO_3$
Temperature   = 16°C
Dissolved residue = 600 mg/L
pH   = 8.7
Determine the saturation index, LSI:
pH_s = $A + B - \log (Ca^{+2}) - \log (\text{alkalinity})$
pH_s = $2.20 + 9.88 - 2.48 - 2.30$
pH_s = 7.3
LSI   = $8.77 - 7.3 = +1.4$
An LSI of +1.4 indicates that this water is scale forming.

COAGULANT EFFECTIVENESS

The removal of suspended solids from surface water is necessary both to make the water aesthetically pleasing to customers and to assist in the elimination of pathogenic organisms. The Surface Water Treatment Rule (SWTR) and the Total Coliform Rule (TCR) require more complete turbidity removal than was previously practiced by most water systems. This requirement in turn demands more efficient coagulation, flocculation, and sedimentation. Effective coagulation is also a tool in removing organic chemical precursors from the raw water.

TABLE 5-2 Threshold odor number corresponding to various dilutions

Volume of Sample (mL) Diluted With Odor-Free Water to 200 mL	Threshold Odor Number
0.8	256
1.6	128
3.1	64
6.3	32
12.5	16
25	8
50	4
100	2
200	1

TABLE 5–3 Constant _A_ as a function of water temperature

Water Temperature, °C	_A_
0	2.60
4	2.50
8	2.40
12	2.30
16	2.20
20	2.10

TABLE 5–4 Constant _B_ as function of total dissolved solids

Total Dissolved Solids, _mg/L_	_B_
0	9.70
100	9.77
200	9.83
400	9.86
800	9.89
1,000	9.90

Significance

Coagulation and flocculation involve the addition of chemical coagulants, such as aluminum sulfate, ferric chloride, or polyelectrolytes, to raw water to hasten the settling of suspended matter. Plant operation is most efficient when the lowest turbidity is obtained in finished water with the lowest cost for coagulant chemicals. Several laboratory tests can provide the information necessary to accomplish this goal. These tests allow operators to select optimal chemical dosages in the laboratory rather than using trial and error in the plant.

The tests can also be used to check the adequacy of flash mixing or flocculation mixing in the plant. Test results may indicate when to improve or modify flash mixers and flocculation basins to obtain more efficient operation. These treatment processes are explained in detail in _Water Treatment_, another text in this series.

When coagulants are evaluated, the goal is to identify the one coagulant, or combination of coagulant and coagulant aids, that will produce low turbidity with the least expensive dose of chemicals. Chemical prices must also be evaluated, because a low dose of an expensive chemical may be more cost-effective than a high dose of an inexpensive chemical.

Sampling

Where the samples are collected for analysis depends on the procedure being utilized and the information desired from the test. The tests should be conducted whenever there is a significant change in water quality or there are other conditions that may require a change in coagulant dose.

Methods of Determination

The following methods are commonly used to determine optimum coagulant effectiveness:

- Jar test
- Zeta potential detector
- Streaming current detector (SCD)
- Particle counting

Jar test

The jar test is readily available to most operators and has been commonly used for many years. There is no standard procedure for conducting the jar test, nor is there standard test equipment that must be used. A typical procedure for conducting a jar test is provided in American Water Works Association (AWWA) Manual M12, *Simplified Procedures for Water Examination.*

Briefly, the procedure is to collect a sample of the raw water and add progressively larger doses of the coagulant chemical being tested to several jars of the sample. The test should be run as quickly as possible after the sample has been collected because a change in water temperature can have a significant effect on the results. Actual plant conditions of flash-mixing speed and time, flocculation time, and settling time are then simulated in the jars.

Visual observation of the floc along with turbidity readings of the settled water in the jars provides the data necessary to make determinations. The types of equipment available for performing jar tests ranges from simple hand-shaken jars to computer-controlled testing devices. A motor-driven jar test device is pictured in Figure 3–18 in chapter 3.

Figure 5–1 illustrates typical jar test results from a water supply in which alum is used as a coagulant. The data show that the benefit of the coagulant decreases with doses larger than 35 mg/L. In other words, beyond that dose it takes a very large increase in the amount of chemical to produce a small increase in turbidity removal. Therefore, 35 mg/L should be considered the optimal dose for alum alone for the water tested.

Zeta potential

Coagulation and flocculation are an electrochemical process in which the electrical resistance between the suspended particles (colloids) in the water is lowered to the point that they will adhere to each other and settle out as a heavy floc.

Many water plants use a zeta meter to assist them in evaluating the effectiveness of coagulant doses. Zeta potential may be viewed as the electrical charge on a suspended particle that allows it to repulse other particles and thus stay suspended. The type and amount of coagulant added reduces the zeta potential, and a zeta meter measures this potential. The closer the reading is to zero, the more the particles will tend to settle, thus indicating more effective coagulation.

Normally, the zeta meter is not an online instrument. Samples may be collected at points in the treatment process to determine coagulant effectiveness, or various coagulant doses may be tested on the raw water in bench-scale experiments. Trained personnel are necessary to operate and interpret the data produced by a zeta meter.

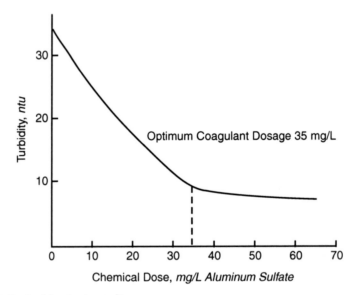

FIGURE 5–1 Typical jar test results

Streaming current detector

An SCD is an online, continuous-monitoring device based on the same electromotive principle as is the zeta meter (Figure 5–2). The detector measures the effectiveness of the coagulant chemical by determining the level of electrical resistance in the treated water after chemical application.

The advantage of having a continuous-monitoring device is that it allows the operator to evaluate changes in the chemical doses as changes in raw-water quality occur. A major concern in installing an SCD is the maintenance (cleaning) of the electrodes and the calibration of the meter to assure accurate readings. Another advantage of the SCD is that it can be used to automatically control coagulant dose by connecting the output signal from the SCD to a coagulant feed pump.

Particle counting

Instruments are now available that enumerate and size the concentration of particles in a water supply (Figure 5–3). These instruments, known as particle counters, combine particle detection technology with electronic counting technology. A sensor detects the particle and converts the information to an electronic signal that is used by the electronic counter.

Particle counters and turbidimeters are similar in that both use a fixed light source to interact with the particles in water. Turbidimeters use light scattering. Particle counters use the principle of light blocking.

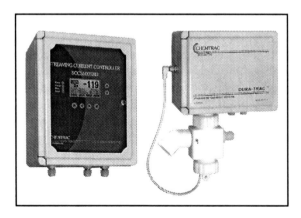

Courtesy of Chemtrac Systems, Inc.

FIGURE 5–2 Streaming current monitor with remote sensor

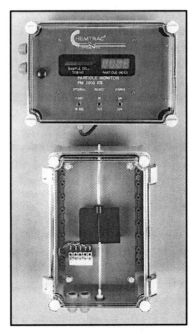

Courtesy of Chemtrac Systems, Inc.

FIGURE 5–3 Liquid particle monitor with remote sensor

Particle counters and turbidimeters are different in that particle counters provide a quantitative measurement, and turbidimeters provide a qualitative measurement. Particle sensors count individual particles according to their size; turbidimeters do not. Particle counters cannot count particles below a given size. Turbidimeters have the ability to detect smaller particles.

Particle counting is a relatively new technique for monitoring filter performance in regard to filter breakthrough and coagulation efficiency and is useful in monitoring pathogen removal. It is used in combination with the pressure hold test to monitor membrane integrity.

As water quality regulations prescribe lower and lower contaminant levels, water treatment plant operators will increasingly have to depend on sophisticated control techniques, such as particle counting, for process control.

COLOR

Color in water may result from the presence of minerals, such as iron and manganese, aquatic organisms, or organic matter from soils and vegetation. A problem in surface waters, some groundwaters containing iron or manganese can also have significant color levels.

Color in water is classified as either true color or apparent color. *True color* is defined as the color of water from which turbidity has been removed. *Apparent color* includes true color and color caused by suspended matter.

Color is determined by a visual comparison of the sample with either a known colored chemical solution or a calibrated color disk.

The unit of measurement is the color unit. Color units are the estimated color of a diluted sample times 50, divided by the milliliters of sample taken for dilution. Color is reported in whole numbers from 1 to 500. Sample pH is always reported with the color units.

Significance

Color in drinking water should be removed to produce a pleasing, acceptable appearance. The color of a drinking water affects consumers' acceptance. Consumers will reject colored water and may change to another source of water even if the other source is less safe. Color may also indicate high levels of organic compounds, which may produce high levels of trihalomethanes (THMs) and other disinfection by-products (DBPs) upon contact with chlorine or other disinfectants.

Color data from raw- and finished-water sample points indicate treatment plant efficiency.

Sampling

Samples collected should represent raw water, finished water, and distribution system water, and should include water pH at the point of sampling. Sampling in surface water systems is especially important.

Methods of Determination

The platinum–cobalt method is the preferred method for color analysis. It is useful for measuring color derived from naturally occurring materials, but it is not applicable to color measurement of waters containing highly colored industrial wastes.

CONDUCTIVITY

Conductivity (specific conductance) is a measure of the ability of an aqueous solution to carry an electric current. This ability depends on the presence of ions; on their total concentration, mobility, and valence; and on the temperature of measurement. Solutions of most inorganic compounds are relatively good conductors. Conversely, molecules of organic compounds that do not dissociate in aqueous solution conduct a current very poorly, if at all.

The units are 1/ohm-m or mho per centimeter. In Système International (SI) units, conductivity is reported as millisiemens per meter (mS/m). To report results in SI units, divide mhos/cm by 10.

Conductivity (specific conductance) is a measure of the ionic strength of a solution. Conductivity is a required water quality parameter of the Lead and Copper Rule. The conductivity of potable waters in the US ranges from 50 to 1,500 µmhos/cm.

Significance

Conductivity is a general parameter that assists the analyst in evaluating many aspects of water quality. In the laboratory, conductivity measurements are used to

- establish the degree of mineralization of a water in order to assess the effect of the total concentration of ions, which is particularly relevant to corrosion rates;
- evaluate variations in the concentration of dissolved minerals in a water source;
- estimate the concentrations of TDS in water supplied to a distribution system or from points in the system;
- approximate the milliequivalents per liter of either cations or anions in a water sample.

Sampling

A 50-mL aliquot of sample for conductivity measurements may be taken from an unpreserved sample collected from a regularly used bacteriological sample point or from a sample collected at the entry point to the distribution system. The sample should be preserved by cooling it to 37°F (4°C) until it is analyzed.

Methods of Determination

Laboratories analyzing a sample for conductivity do not have to be certified, but a US Environmental Protection Agency (USEPA) approved method must be used if the results are to be reported to the state for the Lead and Copper Rule.

An approved method utilizes a self-contained conductivity instrument with conductivity cells containing either platinum or nonplatinum electrodes. Procedures for running the test may be found in *Standard Methods* or *Methods of Chemical Analyses of Water and Wastes* from USEPA. Conductivity measurements may be taken in the field.

HARDNESS

Hardness is a measure of the concentration of calcium and magnesium salts in water, which are generally present as bicarbonate salts. Water hardness is derived largely from the water contacting soil and rock formations. Hard waters usually occur where topsoil is thick and limestone formations are present. Soft waters occur where the topsoil is thin and limestone formations are sparse or absent.

Significance

Hard and soft waters are both satisfactory for human consumption. However, consumers may object to hard water because of the scale it forms in household plumbing fixtures and on cooking utensils. Hardness is also a problem for industrial and commercial users because of scale buildup in boilers and other equipment.

Water that is most satisfactory for household use contains about 75 to 100 mg/L as calcium carbonate ($CaCO_3$). Waters with 300 mg/L as $CaCO_3$ are generally considered too hard. When water is softened in a water treatment plant, it is either partially softened or blended to result in a hardness concentration between 75 and 150 mg/L as $CaCO_3$.

Very soft waters, found in some sections of the United States, have hardness concentrations of 30 mg/L as $CaCO_3$ or less. These waters are generally corrosive and are sometimes treated to increase hardness.

Sampling

If a water treatment plant softens water, hardness analyses of the finished water should be conducted daily to determine whether the desired degree of softening has been achieved. Analyses should be conducted on samples collected immediately after softening and before the water enters a clearwell.

Hardness determinations should also be performed on raw-water samples whenever weather conditions, such as spring rains, affect the supply. This sampling will reveal any variation in the hardness of the raw water and will provide advance information for chemical dosage changes that may be necessary for softening. Even if softening is not practiced, hardness determinations should still be made periodically as a general water quality measurement.

At least 100 mL of sample should be collected in either glass or plastic bottles. Samples may be stored for no longer than 7 days before analysis. They should be cooled to 39°F (4°C) and acidified with 0.5 mL/100 mL of nitric acid unless they are going to be analyzed immediately.

Methods of Determination

The ethylenediaminetetraacetic acid (EDTA) titrimetric method is the preferred method of analysis. It consists of sequestering (tying up) the calcium and magnesium ions by titrating with an EDTA solution. The sample is titrated in the presence of an indicator. The initial solution is red, and it changes to blue when all the ions have been sequestered.

TASTE AND ODOR

Tastes and odors in water are difficult to measure. They are caused by a variety of substances, including organic matter, dissolved gases, and industrial wastes. Odors in water supplies are most frequently caused by algae or decaying organic matter. The intensity and offensiveness of odors vary with the type of organic matter.

Odors are generally classified as aromatic, fishy, grassy, musty, septic, or medicinal. Industrial wastes, such as phenolic or oil waste, are also responsible for some odors in surface waters.

The human sense of smell is much more sensitive than the sense of taste, so odor tests are most commonly run in water treatment plants. The taste test, which classifies tastes as sweet, sour, bitter, and salty, can only be run on water known to be safe for drinking; thus its usefulness is limited.

Significance

An odor test can be used to evaluate how well a water treatment plant removes taste- and odor-causing organic materials. The test can also be used to trace the origins of contamination in source water. For example, an odor-causing industrial waste discharge might be occurring upstream from a water treatment plant. Samples can be collected at intervals upstream until the problem-causing area has been reached. The odors should become stronger closer to the discharge and should not be evident in samples collected upstream from the discharge. This technique is time consuming; however, it can be conducted by water plant personnel, and extensive laboratory facilities are not necessary. An odor test can also be used to detect problems in the distribution system. For example, odors will occur in dead-end water mains having a significant bacteriological buildup. A definite chlorine odor can indicate the loss of free chlorine caused by stagnation, slime buildup, and/or anaerobic conditions.

The threshold odor number (TON) is designed to help monitor all types of odors, independent of source. The TON cannot, however, be used to indicate the concentration of the odor-producing substance because some substances produce strong odors at low concentrations. For example, some chemical wastes, such as phenol in chlorinated water, have been detected by the threshold odor test at a

0.001-mg/L concentration. Other odor-producing substances, such as detergents, may not be detected until the concentration is as high as 2.5 mg/L.

An odor with a TON of 3 might be detected by a consumer whose attention is called to it, but it probably would not be noticed otherwise. If an odor appears gradually, consumers will adapt to it, and it will be noticed less than if it appears suddenly. Finished-water quality with a TON above 5 will begin to draw complaints from consumers. When a TON of 3 or more is detected in a finished-water supply, quick action should be taken to solve the problem.

Sampling

Water supplies with seasonal or recurring taste-and-odor problems should be analyzed regularly and, as problems occur, corrective action should be taken. The tests may be time consuming, so it is not generally possible to conduct more than a few tests per day.

Water samples should be taken from raw and finished waters. At least 1,000 mL of sample should be collected for an odor analysis. Samples should be collected in clean bottles that have not been used for any samples that might leave a taste or an odor. The bottles should be washed with detergent and rinsed with distilled water. Glass sample bottles should be used because plastic containers may add some odor of their own or an odor from substances that were previously in the container.

Aeration and mixing of the sample should be kept to a minimum before testing because air will strip or oxidize odor-producing compounds. An air space should be left at the top of the bottle so that the sample can be thoroughly shaken before testing.

Odor tests should be run as soon as possible after collection. If the sample must be stored, it should be tightly capped and placed in an odor-free refrigerator. The sample should be analyzed no later than 24 hours after collection.

Methods of Determination

Two standard methods of quantifying taste and odor in water supplies are in use: the threshold odor test and the flavor profile analysis.

Threshold odor test

Most tastes and odors are extremely complex, and the best way to detect them is with the human sense of smell. A series of sample dilutions are prepared and placed in bottles for observers to test. Each bottle contains 200 mL of liquid consisting of a mixture of sample and odor-free distilled water.

The bottles are arranged so that the observer smells the most dilute samples first and then smells from a bottle of completely odor-free water as a reference. Then the next strongest sample is smelled, and again the odor-free water, and the process is continued until an odor is first detected. The TON may be calculated as follows:

$$\frac{V_s + V_d}{V_s} = \text{TON}$$

Where:

V_s = volume of sample, in mL
V_d = volume of dilution water, in mL

The lowest obtainable TON is 1. If no odor is detected in an undiluted sample, the TON is reported as "no odor observed" and no number is assigned. If an odor is first detected in a bottle that has 100 mL of sample diluted to 200 mL with distilled water, the TON is 2. TONs corresponding to various dilutions are shown in Table 5–2.

The threshold odor test is not precise and is based on human judgment. The ability to detect odors varies among individuals, and if very accurate results are desired, a panel of five or more persons is recommended to overcome this variability. Persons performing odor tests should not have colds or allergies that would affect their sense of smell. They also should be nonsmokers because smoking dulls the sense of smell. Plant operators should not make odor observations because they work with the water all day and their sense of smell is used to it. All tests must, of course, be conducted in an odor-free atmosphere.

Flavor profile analysis

The flavor profile analysis (FPA) was approved by the Standard Methods Committee in 1990 and first appeared as a proposed standard method in the eighteenth edition of *Standard Methods*.

The FPA differs from the TON in that the samples are not diluted and each individual odorant in the sample is evaluated and numerically rated. FPA can be applied to both taste- and odor-causing compounds.

In the procedure, a panel consisting of four to six members and a panel coordinator conduct each round of testing. Panel members must be selected for their desire to participate, their tested ability to accurately taste and smell samples, and their ability to interact with other panel members. (Dominant personalities are not desirable.)

The panel members gather in an odor-free room. The panel coordinator prepares and presents the samples to the panel members, and they independently write down the taste or odor attributes they have observed. The coordinator writes down each observation on a blackboard or flip chart and leads a discussion in order to reach a consensus among the panel members.

Panelists are trained in the proper methods of tasting and sniffing samples and are taught to identify and rate the attributes of both tastes and odors. They must also be trained in how to prepare for a round of testing, i.e., no aftershave or perfume, no smoking or gum chewing 1 hour prior to the test, no colds or allergies.

Standards and references for tastes and odors are commercially available to assist in training panelists and reaching consensus. For example, to help panel members identify the four taste attributes, standards can be made from sucrose for sweetness, citric acid for sourness, sodium chloride for saltiness, and coffee for bitterness. Natural material is also available for panel use; for example, geranium leaves may be used to assist the panel in coming to a consensus on what constitutes a geranium odor in a water sample.

TEMPERATURE

Temperature is measured on either the Fahrenheit (°F) or the Celsius (°C) scale. The freezing point of water is 32°F or 0°C; the boiling point is 212°F or 100°C. Because temperature is a factor in computing the Langelier saturation index, it is one of the water quality parameters required by the Lead and Copper Rule.

Significance

Water temperature determines, in part, how efficiently certain unit processes operate in the treatment plant. The rate at which chemicals dissolve and react is somewhat dependent on temperature. Cold water generally requires more chemicals for efficient coagulation and flocculation to take place. Water with a higher temperature may result in a higher chlorine demand because of increased reactivity and also because there is usually an increased level of organic matter, such as algae, in the raw water.

Sampling

Temperature readings must be taken onsite, either directly from the water or from samples immediately after collection. Immediate readings are necessary because the water temperature will begin to change once the sample is taken.

Methods of Determination

A laboratory thermometer is used for temperature analysis. The thermometer is left in the water long enough to get a constant reading, and the measured temperature is expressed to the nearest degree or less depending on the thermometer's accuracy. Digital battery-operated thermometers are currently available. They have the advantage over glass bulb–type thermometers of being easier to read and less prone to break.

TOTAL DISSOLVED SOLIDS

Total dissolved solids (TDS), also referred to as *total filterable residue*, in natural waters consists mainly of carbonates, bicarbonates, chloride, sulfate, calcium, magnesium, sodium, and potassium. Dissolved metals, dissolved organic matter, and other substances also account for a small portion of the dissolved residue in water.

Significance

Dissolved solids in drinking water tend to change the water's physical and chemical nature. Distilled or deionized water has a flat taste; water with some dissolved solids is preferred by most consumers. Different salts in solution may interact and cause effects that each salt alone would not cause. The presence of harmful dissolved compounds or ions (such as arsenic and mercury) can be dangerous in water even where the total solids concentration is relatively low.

It is generally agreed that the TDS concentration of palatable water should not exceed 500 mg/L. Lime softening and ion exchange facilities both significantly reduce the quantity of TDS in finished water.

Many communities in the United States use waters containing 2,000 mg/L or more TDS because better-quality water is not available. These waters tend to be unpalatable, may not quench thirst, and can have a laxative effect on new or transient users. However, no lasting harmful effects have been reported from such waters. Waters containing more than 4,000 mg/L TDS are considered unfit for human consumption.

Raw-water source samples can be dipped from just below the surface of the water in the area of the intake structure. They may be collected in clean wide-mouth glass or plastic 1-L containers. Filter effluent samples should be collected from an effluent sample tap or drain line. The sample should be stored in a cooler away from sunlight for transportation to the laboratory.

TURBIDITY

Turbidity is an optical property caused by particles suspended in water. These particles cause light rays to be scattered and absorbed rather than to be transmitted in a straight line, making the water appear cloudy. Turbidity is a method of measuring the clarity of a water sample. Waters showing very little light scattering produce low-turbidity measurements; those with a great deal of light scattering indicate high turbidity. The suspended particles causing turbidity include organic and inorganic matter and plankton.

Turbidity should not be confused with suspended solids. Turbidity expresses how much light is scattered by the sample. Suspended solids measurements express the weight of suspended material in the sample. In most cases, turbidity cannot be correlated to suspended solids concentration or any quantitative analysis.

Interferences with turbidity measurement include color, high turbidity, and gas bubbles. Scratched sample cells, condensation, poor calibration, and stray light also interfere.

Significance

Turbidity is expressed in nephelometric turbidity units (ntu), and a reading in excess of 5 ntu is noticeable to consumers. Turbidity is significant in water supplies because it creates a potential public health hazard, unpleasant appearance, and operational difficulties. The most important of these is the potential public health hazard. The effectiveness of chlorine or other disinfectants depends on the disinfectant making contact with the pathogenic organisms in the water. Suspended particles in turbid water can shelter microorganisms from the disinfectant and allow them to still be viable when they reach the customers.

Turbid water may also contain particles of organic matter that can react with chlorine to form THMs or other DBPs.

Turbidity analyses are also used to evaluate in-plant operations. Turbidity measurements taken after settling and before filtration serve to monitor the performance of the coagulation, flocculation, and sedimentation processes. A rise in turbidity after settling indicates that the coagulant application should be changed and/or that operational corrections must be made.

Settled water before filtration should have a turbidity of less than 10 ntu. If water with high turbidity reaches the filter, it will cause high filter head loss and shorten filter runs. Changes in raw-water turbidity usually require that the coagulant dose be changed. Any noticeable change in turbidity within the unit process should be an immediate warning that operational adjustments are necessary.

Turbidity analyses are also used to monitor finished-water quality for compliance with state and federal drinking water standards.

Sampling

Turbidity analysis is usually conducted on samples collected from raw water, sedimentation basin effluent, filter effluent, and finished water. Figure 5–4 shows some typical turbidity sampling points.

At least 100 mL of sample should be collected in a clean glass or polyethylene container. Samples should be shaken and analyzed immediately after collection because the level of turbidity can change if the sample is stored. If it is not possible to run a turbidity test immediately, the sample should be stored in the dark for no longer than 24 hours.

All filter plants should keep a continuous record of finished-water turbidity. Continuous-reading turbidimeters with recorders installed on the filter-effluent piping will continuously determine, report, and record the quality of the filter effluent. The turbidimeter signal can sound alarms to indicate the need to shut down an improperly operating filter. This alarm system increases the reliability of the filter operation and is especially important in assuring the safe operation of pressure filters and high-rate (4 to 6 gpm/ft^2 [2.7 to 4.1 mm/sec]) filter plants.

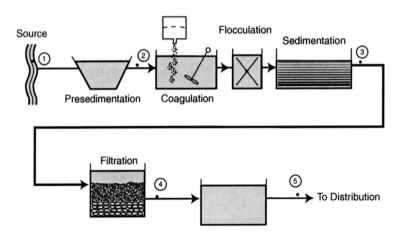

1. Turbidity of raw water entering the plant.
2. Turbidity reduction by presedimentation; helps the operator determine coagulant dose.
3. Turbidity removal by coagulation, flocculation, and sedimentation processes; assists the operator in monitoring the efficiency of the process.
4. Turbidity after filtration; continuous monitoring of turbidity for each filter monitors for turbidity breakthrough, which is one of the indicators of the need for filter backwashing.
5. Turbidity of all treated water leaving the plant; monitors compliance with drinking water regulations for maximum allowable effluent turbidity.

FIGURE 5–4 Typical turbidity sampling points

Methods of Determination

The nephelometric turbidimeter measures the scattering of light in nephelometric turbidity units. The USEPA drinking water regulations specify the use of a nephelometric turbidimeter for all required monitoring.

Analysis is quick and easy with the nephelometric method. Nephelometry is useful for in-plant monitoring, and results can be compared from plant to plant, which is an advantage to operators seeking performance information from other facilities.

Under the requirements of the SWTR, water systems serving a population in excess of 500 must perform turbidity monitoring of filtered water at least every 4 hours that the plant is in operation. A system may substitute continuous turbidity monitoring if the equipment is validated on a regular basis using a procedure approved by the state.

SELECTED SUPPLEMENTARY READINGS

Gordon, G., W.J. Cooper, R.G. Rice, and G.E. Pacey. 1992. *Disinfection Residual Measurement Methods.* 2nd ed. Denver, Colo.: American Water Works Association Research Foundation and American Water Works Association.

Jensen, J.N., and J.D. Johnson. 1989. Specificity of the DPD and Amperometric Titration Methods for Free Available Chlorine. *Jour. AWWA,* 81(12):59.

Mallevialle, J., and I.H. Suffet, eds. 1987. *Identification and Treatment of Tastes and Odors in Drinking Water.* Denver, Colo.: American Water Works Association Research Foundation and American Water Works Association.

Manual M12, Simplified Procedures for Water Examination. 2001. Denver, Colo.: American Water Works Association.

Manual of Instruction for Water Treatment Plant Operators. 1975. Albany, N.Y.: New York State Department of Health.

Manual of Water Utility Operations. 8th ed. 1988. Austin, Texas: Texas Water Utilities Association.

Meng, A.-K., and I.H. Suffet. 1992. Assessing the Quality of Flavor Profile Analyses. *Jour. AWWA,* 84(6):89.

Methods of Chemical Analyses of Water and Wastes. 1984. EPA-600/4-74-020. Cincinnati, Ohio: US Environmental Protection Agency.

Standard Methods for the Examination of Water and Wastewater. 20th ed. 1998. Washington, D.C.: American Public Health Association, American Water Works Association, and Water Environment Federation.

Suffet, I.H., J. Mallevaille, and E. Kawczynski. 1995. *Advances in Taste-and-Odor Treatment and Control.* Denver, Colo.: American Water Works Association Research Foundation and American Water Works Association.

6

Inorganic Chemicals

An inorganic chemical, substance, compound, or contaminant is one not derived from hydrocarbons. An inorganic contaminant is an inorganic substance regulated by the USEPA. Inorganic contaminants contained in this list are antimony, asbestos, barium, beryllium, cadmium, chromium (total), copper, cyanide, fluoride, lead, mercury, nickel, nitrate, nitrite, selenium, thallium, and arsenic.

In drinking water, other inorganic compounds are of interest. These include the divalent cation, calcium, iron, and manganese.

CARBON DIOXIDE

Carbon dioxide is a colorless, odorless, noncombustible gas that is found in all natural waters. Carbon dioxide in surface waters can originate from the atmosphere, but most comes from biological oxidation of organic matter. Biological oxidation is also the primary source of carbon dioxide in groundwater.

Significance

Consuming excess carbon dioxide in water has not been found to have adverse public health effects. In fact, carbon dioxide is present in commercial carbonated beverages in concentrations far greater than those found in natural waters. However, carbon dioxide in water can cause corrosion problems. In addition, carbon dioxide values must be known to calculate proper lime dosages when softening water. If recarbonation is used following lime softening, carbon dioxide values must be determined in order to control the process.

If recarbonation is used following lime softening, carbon dioxide values must be determined in order to control the process.

145

Sampling

Carbon dioxide analyses should be run on raw and finished water. Special precautions must be taken during collection and handling of the sample if the titrimetric method is being used. Exposure to the air must be kept to a minimum. Field determination of free carbon dioxide immediately after sampling is advisable. If field determination is impossible, the sample should be kept cool and the analysis completed as soon as possible.

Samples may be collected in glass or plastic bottles. At least 100 mL of sample should be collected. The bottle should be filled to the top with no air space left, and no preservatives should be added.

Methods of Determination

The amount of carbon dioxide in water may be determined by using the nomographic method or the titrimetric method. In order to use the nomographic method, the pH, bicarbonate alkalinity, temperature, and total dissolved solids must be known. Results are most accurate when the pH and alkalinity are analyzed immediately after sample collection. The titration method may be performed potentiometrically or with phenolphthalein indicator.

CHLORINE RESIDUAL AND DEMAND

Chlorine is usually added to source water as it enters the water treatment plant (prechlorination) and again just before it leaves the plant (postchlorination). Postchlorination is primarily administered to provide an excess of chlorine for continued disinfection in the distribution system. Chlorine is also often added at intermediate points during the treatment process. Tests of chlorine levels in the plant and throughout the distribution system are necessary to determine that chlorine dosage levels are adequate and to monitor water quality.

Significance

Destruction of pathogenic organisms by chlorine is directly related to contact time and the concentration of the chlorine. High chlorine doses with short contact periods will provide essentially the same results as low doses with long contact periods. Chlorination also oxidizes substances, such as iron, manganese, and organic compounds, making their removal from the water possible.

Successful chlorination requires that enough chlorine be added to complete the disinfection or oxidation process. However, chlorine must not be added in amounts that are wasteful, creating unnecessarily high operational costs. Determining effective and efficient chlorine dosage levels is the responsibility of the plant operator.

146

Chlorine Residual

There are two types of chlorine residual: combined residual and free available residual. The process by which these are formed is illustrated in Figure 6–1.

The first amount of chlorine (for example, 1 mg/L) that is added to raw water is destroyed by reducing compounds, such as iron and manganese (from point 1 to point 2 in the figure). The chlorine oxidizes the iron and manganese and in the process is used up—no residual forms and no disinfection occurs.

If the initial chlorine dosage is higher (for example, 2.5 mg/L), the reaction will go to point 3. Between points 2 and 3, the chlorine reacts with the organic substances and the ammonia in the water, forming chloroorganics and chloramines. These two products are called *combined chlorine residual*. This is a chlorine residual that, because it has combined with other chemicals in the water, has lost some of its disinfecting strength. Compared to free chlorine, combined chlorine residual has relatively poor disinfecting power and may cause tastes and odors characteristic of water in a swimming pool.

As the chlorine dosage is increased further (point 3 to point 4), the chloramines and some of the chloroorganics are destroyed. This process reduces the combined chlorine residual until, at point 4, the combined residual reaches its lowest point. Point 4 is called the *breakpoint*. At the breakpoint, the chlorine residual changes from combined to free available.

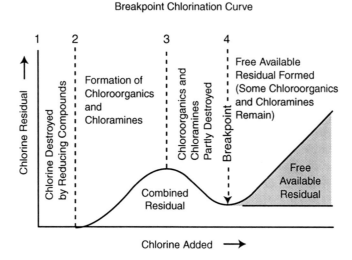

FIGURE 6–1 Formation of combined chlorine residual and free available chlorine residual

As the initial chlorine dosage is increased still further (beyond 4 mg/L in this example), free available chlorine residual is formed—"free" in the sense that it has not reacted with anything and "available" in the sense that it can and will react if necessary. In terms of disinfecting power, free available residual is 25 times more powerful than combined residual, and it will not produce the characteristic swimming-pool odor that combined residuals do. Because free available chlorine residual forms only after the breakpoint, the process is called *breakpoint chlorination*.

The free available chlorine residual at the consumer's tap should be at least 0.2 mg/L or at a level specified by the state. This level helps ensure that the water is free from harmful bacteria. However, higher levels may be necessary to control special problems, such as iron bacteria. If maintaining a free chlorine residual in a distribution system becomes difficult, several possible problems are indicated. Stagnant water in dead ends, biological growths, contamination of mains during main-break repair, and contamination caused by cross-connections all cause dissipation of free chlorine residuals. Further, a drop in chlorine residual in the distribution system may indicate inadequacies in the treatment process itself.

Chlorine Demand

Test results for chlorine residual can be combined with operating data regarding the amount of chlorine added at the plant to yield information on chlorine demand. Chlorine demand is a measurement of how much chlorine must be added to the water to achieve breakpoint chlorination or whatever free chlorine residual is desired.

The most significant reason for analyzing a water supply's chlorine demand is to determine the proper dosage. However, changes in chlorine demand can also indicate water quality changes. For example, if a water supply suddenly requires more chlorine in order to maintain a residual (that is, if the water exhibits a higher chlorine demand), then the chlorine is oxidizing some contaminants that previously were not present in the water supply.

When chlorine demand increases, two steps are necessary. First, the chlorine dose must be increased to meet the higher demand. Second, the reason for the increased demand should be investigated. A sudden increase in chlorine demand frequently occurs in surface water because of seasonal water quality changes. Chlorine demand in groundwater should not change substantially because the quality of groundwater is usually very stable.

DISINFECTION BY-PRODUCTS

The first disinfection by-products (DBPs) to be regulated were the trihalomethanes. Continuing studies and research have revealed that chlorine (and all other alternate disinfectants) reacts with organic compound precursors in the water to form many different kinds of organic compounds. Many of these compounds are considered to be toxic or carcinogenic. Haloacetic acids, halonitriles, haloaldehydes, and chlorophenols are just a few of the organic compounds associated with chlorine disinfection. Thus, chlorination has its good and its bad points; the water plant operator must know how to adequately disinfect the water without producing undesirable levels of DBPs.

Sampling

Chlorine residual sampling is done at the treatment plant and at the consumer's faucet.

Treatment plant sampling

In-plant sampling of chlorine residual determines whether sufficient chlorine has been added to the water before it leaves the treatment plant. This is the only way to be sure that finished water leaving the plant contains the desired chlorine residual. Obtaining representative samples is the most critical part of in-plant chlorine sampling.

In some instances, a sample must be collected at a point near the location of chlorine addition; in which case, analyses will probably show a disproportionately high chlorine residual. To obtain data that approximate the actual chlorine residual in a particular basin, the sample should be held for a time period equal to the basin detention time or, in any case, at least 10 minutes. Under requirements of the Surface Water Treatment Rule, surface water systems serving populations larger than 3,300 are required to provide continuous chlorine residual monitoring where the water enters the distribution system.

Distribution system sampling

Sampling for chlorine residual from the distribution system is done to determine whether consumers are receiving water that is safe to drink. In other words, if there is no chlorine residual, live pathogenic organisms could be present. The chlorine residual analyses, if made at the same time and location as are analyses of bacteria samples, may be correlated with coliform results.

If analysis is made in the field, only about 10 mL of sample are required. If samples are to be taken back to a laboratory, a 100-mL sample should be collected. Analysis should be completed as soon as possible after collection. Preservation is not recommended because chlorine is unstable in water and residual chlorine will continue to diminish with time.

Agitation or aeration of the sample should be avoided because it can cause reduction of the sample's chlorine concentration. Chlorine will also be destroyed and subsequent analysis will be erroneously low if samples are exposed to sunlight. The same sample bottle should never be used for both chlorine residual and coliform analyses. Bottles used for coliform analysis contain a chemical (sodium thiosulfate) that neutralizes the chlorine residual.

Methods of Determination

The *N,N*-diethyl-*p*-phenylenediamine (DPD) color comparator test kit is the simplest and quickest way to test for residual chlorine. The test takes approximately 5 minutes to complete. The orthotolidine method should not be used because it is not as accurate as the DPD method, particularly for measuring free chlorine residual.

Another technique, used primarily in laboratories because of its accuracy, is amperometric titration. The method is unaffected by sample color or turbidity, which can interfere with colorimetric determinations. However, performance of amperometric titration requires greater skill and care than does the DPD method.

The chlorine demand can be determined by treating a series of water samples with known but varying chlorine dosages. After an appropriate contact time, the chlorine residual of each sample is determined. This procedure will indicate which dosage satisfied the demand and provided the desired residual.

DISSOLVED OXYGEN

Dissolved oxygen (DO) in water is not considered a contaminant. An excess or a lack of DO does, however, help create unfavorable conditions. Generally, a lack of DO in natural waters creates the most problems, specifically an increase in tastes and odors as a result of anaerobic decomposition.

The amount of DO in water is a function of the water's temperature and salinity. Cold water contains more DO, and saline water contains less DO. Natural waters are seldom in equilibrium (exactly saturated with DO). Temperature changes as well as chemical and biological activities all use or release oxygen, causing the amount of DO in water to change continually.

Significance

Nominal levels of dissolved oxygen in municipal water supplies are generally not a problem. DO has no adverse health effects and actually increases water's palatability. Most consumers prefer water that has a DO content near the saturation point. However, a concentration this high is detrimental to metal pipes because oxygen helps accelerate corrosion.

Introducing oxygen into water can be a method of treatment for purposes such as oxidizing iron and manganese into forms that will precipitate out of the water. DO also removes excess carbon dioxide. Provided the contact time is long enough, DO will also help degrade some organic compounds that cause taste-and-odor problems.

Where additional DO is desired for water treatment, some form of aeration is used. Various types of aeration processes are detailed in *Water Treatment*, another book in this series.

DO data from raw-water storage-reservoir samples can also be used to indicate the general quality of the water. On the basis of these data, operators may be able to make treatment changes or alter the way the reservoir releases are made in order to prevent taste, odor, and other problems.

Sampling

DO analyses should be conducted routinely on raw-water samples from surface water sources, particularly if storage reservoirs are being used. Treated-water samples should also be analyzed routinely if aeration is used as a treatment process; otherwise the tests may be conducted on a weekly basis for general quality data.

DO should be determined onsite if the electrode method (see next section) is used. If the modified Winkler test (see next section) is used, the sample must be collected in a glass bottle and "fixed" (treated with a chemical additive to retard change) onsite. The sample should be stored in the dark at the temperature of the collection water or water sealed and kept at a temperature of 50 to 68°F (10 to 20°C) until the analysis can be performed.

Methods of Determination

The electrode method and the modified Winkler method (also called the *iodometric method*) are preferred for DO measurements. Because the electrode method is not as sensitive to interferences as is the modified Winkler test, it is excellent for analyzing DO in polluted waters, highly colored waters, and strong waste effluents. Drinking waters and supply reservoirs have few interferences that cause problems with the modified Winkler procedure.

INORGANIC METALS

The effects of metal in water are varied. Table 1–1 in chapter 1 provides information on inorganic contaminants including metals covered under the National Primary Drinking Water Regulations, the MCLG, MCL, or treatment technique. Potential health effects from ingestion of water, sources, and treatment technology information is provided.

Three types of metals are discussed in this section. Dissolved metals are metals in an unacidified sample that pass through a 0.45 μm membrane. Suspended metals are metals in an unacidified sample retained by a 0.45 μm membrane. Total

metals are the sum of suspended and dissolved metals or the concentration determined in an unfiltered digested sample.

Sampling and Sample Preservation

Before collecting a sample, it must be decided what fraction is to be analyzed (dissolved, suspended, total, or acid-extractable). This decision will determine in part
whether the sample is acidified with or without filtration and the type of digestion
required.

Serious errors may be introduced during sampling and storage because of contamination from the sampling device, failure to remove residues of previous samples from the sample container, and loss of metals by adsorption on or
precipitation in the sample container caused by failure to acidify the sample properly.

Sample containers

The best sample containers are made of quartz or PTFE. Because these containers
are expensive, the preferred sample container is made of polypropylene or linear
polyethylene with a polyethylene cap. Borosilicate glass containers also may be
used, but soft glass containers should be avoided for samples containing metals in
the microgram-per-liter range. Samples should be stored for determination of sil-
ver in light-absorbing containers. Only containers and filters that have been acid
rinsed should be used.

Preservation

The samples should be preserved immediately after sampling by acidifying with
concentrated nitric acid (HNO_3) to pH < 2. The samples should be filtered for dissolved metals before preserving. Usually 1.5 mL concentrated HNO_3/L sample (or
3 mL 1 + 1 HNO_3/L sample) is sufficient for short-term preservation. For samples
with high buffer capacity, the amount of acid should be increased (5 mL may be
required for some alkaline or highly buffered samples). Commercially available
high-purity acid[*] should be used or high-purity acid should be prepared by sub-
boiling distillation of acid.

After acidifying the sample, it is preferable to store it in a refrigerator at approximately 4°C to prevent change in volume caused by evaporation. Under these
conditions, samples with metal concentrations of several milligrams per liter are
stable for up to 6 months (except mercury, which is stable up to 5 weeks). For

---

*Ultrex, J.T. Baker, or equivalent.

microgram-per-liter metal levels, the samples should be analyzed as soon as possible after sample collection.

Methods of Determination

The presence of metals is determined using both colorimetric and instrumental methods. Instrumental methods include atomic absorption spectrometry, which includes flame, electrothermal (furnace), hydride, and cold-vapor techniques; flame photometry; inductively coupled plasma emission spectrometry; inductively coupled plasma mass spectrometry; and anodic stripping voltametry.

Sources of Contamination

Introducing contaminating metals from containers, distilled water, or membrane filters should be avoided. Some plastic caps or cap liners may introduce metal contamination; for example, zinc has been found in black bakelite-type screw caps as well as in many rubber and plastic products; and cadmium has been found in plastic pipette tips. Lead is a ubiquitous contaminant in urban air and dust.

Take care not to introduce metals into samples during preliminary treatment. During pretreatment, contact with rubber, metal-based paints, cigarette smoke, paper tissues, and all metal products including those made of stainless steel, galvanized metal, and brass should be avoided. Conventional fume hoods can contribute significantly to sample contamination, particularly during acid digestion in open containers. Vessels should be covered with watch glasses and spouts turned away from incoming air to reduce airborne contamination. Plastic pipette tips often are contaminated with copper, iron, zinc, and cadmium. Colored plastics, which can contain metals, should not be used. Certified metal-free plastic containers and pipette tips should be used when possible. Glass should not be used if analyzing for aluminum or silica.

Metal-free water should be used for all operations. Reagent-grade acids used for preservation, extraction, and digestion should be pure. If excessive metal concentrations are found, the acids should be purified by distillation or ultrapure acids should be used. Blanks should be processed through all digestion and filtration steps, and blank results should be evaluated relative to corresponding sample results. Either corrections should be applied to sample results or other corrective actions should be taken as necessary or appropriate.

Inorganic Nonmetallic Compounds

Inorganic nonmetallic contaminants include cyanide, fluoride, nitrate, and nitrite. The analytical methods used for determination include wet-chemical techniques and instrumental methods, such as ion chromatography.

Nitrate generally occurs in trace quantities in surface water and in high concentrations in some groundwaters. Nitrite is an intermediate product formed when ammonia is oxidized to nitrate and when nitrate is reduced. Cyanide is not normally found in either surface water or groundwater. Fluoride, iron, and manganese are discussed in detail in the following sections.

FLUORIDE

Fluoride is found naturally in many waters. It is also added to many water systems to reduce tooth decay.

Significance

Research has demonstrated that drinking water containing a proper amount of fluoride results in a 65 percent average reduction in tooth decay during the years of children's tooth formation (from birth to between the ages of 12 and 15).

Fluoride concentrations in drinking water that are optimum for reducing tooth decay vary with climate. Because more water is consumed in warmer climates, fluoride concentrations should be lower in these areas. Excessive fluoride concentrations can cause teeth to become stained or mottled. This problem generally occurs only where natural fluoride concentrations exceed 2.0 mg/L.

Close control of fluoride concentrations is necessary to assure the maximum benefit of fluoridation with an adequate margin of safety. A reduction of only 0.3 mg/L below the optimum concentration can drastically reduce the dental benefits of fluoride. Specific guidelines from the state should be obtained concerning recommended concentrations for a given water supply if fluoride is being added to the water.

Sampling

Fluoride samples should be collected from raw and finished water in polyethylene bottles. Raw-water samples are necessary because the total amount of fluoride reaching the consumer is equal to the fluoride concentration in the raw water plus the fluoride added at the plant. Although the fluoride level in most source water is fairly stable, it can vary somewhat and so should be periodically analyzed. The amount of fluoride to be added to the raw water is calculated by subtracting the raw-water concentration from the desired treated-water concentration.

Finished-water samples are tested to ensure that the fluoride feeders are operating correctly and the final fluoride concentration is at the desired level. Samples should be tested daily.

Samples collected for fluoride analysis may be held for 7 days before analysis. They should be stored in a refrigerator at 39°F (4°C) with no preservatives added.

Methods of Determination

Two methods for fluoride analysis are commonly used: the SPADNS (sodium,2-(parasulfophenylazo)-1.8-dihydroxy-3,6-naphthalene disulfonate) method and the electrode method. The electrode method requires a selective ion fluoride electrode connected to a pH meter with a millivolt scale or to a meter having a direct concentration scale for fluoride. With either method, interferences may require a distillation step prior to the test. The water to be tested should be checked for interferences that might be present.

IRON

Iron occurs naturally in rocks and soils and is one of the most abundant of all elements. It exists in two forms: ferrous (Fe^{+2}) and ferric (Fe^{+3}). Ferrous iron is found in well waters and in waters with a low level of dissolved oxygen. Under anaerobic conditions, waters can have significant dissolved-iron concentrations.

Dissolved iron in water is derived naturally from soils and rocks. It may also result from the corrosive action of water on unprotected iron or steel mains, steel well casings, and tanks. Surface waters may also occasionally contain appreciable amounts of iron that originate from industrial wastes or from acid runoff from mining operations.

Significance

There are no known harmful health effects from drinking water containing iron. Water quality limits on allowable concentrations of iron in water supplies are based on the problem of discoloration and undesirable taste and of the iron staining porcelain plumbing fixtures. Iron concentrations above 0.3 mg/L can cause undesirable "red water." Concentrations at or above this level in finished water indicate that steps should be taken to provide iron removal. Details on iron removal and control methods are detailed in *Water Treatment*, another book in this series.

Iron also provides a nutrient source for some bacteria that grow in distribution systems and wells. Iron bacteria, such as *Gallionella*, cause red water, tastes and odors, clogged pipes, and pump failure.

Whenever tests show increased iron concentrations between the water plant and the consumer's tap, corrosion and/or iron bacteria may be present, and corrective action should be taken. If the water is corrosive, pH adjustment might be considered first. If the problem is caused by bacteria, flushing of the mains, shock chlorination (temporarily high concentrations), or increased everyday chlorination may prove effective.

Sampling

Samples should be taken from raw and finished water. The samples should be collected in glass or plastic bottles and may be stored as long as 6 months before analysis. At least 100 mL of sample should be collected. Samples should be preserved with approximately 0.5 mL of concentrated nitric acid/100-mL sample to lower the pH to less than 2.

Methods of Determination

Iron concentration may be determined by the phenanthroline method or the atomic absorption spectrophotometric (AA) method. The phenanthroline method is simple and reliable. It is a colorimetric test and can be run with a spectrophotometer or filter photometer. The AA method, used by large laboratories, is very accurate and is particularly advantageous when large numbers of samples must be tested.

MANGANESE

Manganese, a metal, creates problems in a water supply similar to those created by iron. Manganese occurs naturally in ores but not in a pure state. It exists in soils primarily as manganese dioxide. It is found both in the manganous divalent form (Mn^{+2}) and in the quadrivalent form (Mn^{+4}). Manganese is much less abundant in nature than is iron; therefore, it is found less often in water supplies and is present at lower concentrations. It is also more difficult than iron is to oxidize or cause to precipitate because manganous solutions are more stable than are ferrous solutions.

The most common forms of manganese—oxides, carbonates, and hydroxides—are only slightly soluble. Consequently, manganese concentrations in surface waters seldom exceed 1.0 mg/L. In groundwaters subject to anaerobic or reducing conditions, manganese concentrations, like iron concentrations, can become very high.

Significance

Consumption of manganese has no known harmful effects on humans. Water quality limits on allowable concentrations of manganese have been based on aesthetic problems rather than health concerns. Manganese does not usually discolor water, but it stains clothes and bathroom fixtures black. Staining problems begin at 0.05 mg/L, a much lower concentration than for iron.

Raw-water and finished-water analyses will indicate whether manganese removal is necessary or whether the desired removal has been achieved in the treatment plant. Increases in manganese concentration in the distribution system are not generally experienced, except that a rapid flow change in the distribution system may result in some deposits breaking loose and entering consumers' water. This problem is best controlled by flushing the lines in areas where the problem occurs.

Sampling

Samples for manganese analysis should be taken from raw and finished water. The samples should be collected in glass or plastic bottles and may be stored as long as 6 months before analysis. At least 100 mL of sample should be collected. Samples should be preserved with concentrated nitric acid. Approximately 0.5 mL of concentrated nitric acid/100-mL sample should be added to lower the pH to less than 2.

Methods of Determination

The AA method is the preferred method of determination.

pH

pH is a measure of the hydrogen ion concentration present in water; or the logarithm of the reciprocal of the hydrogen ion concentration or the negative $\log_{10}$ of the hydrogen ion concentration in water.

$$\begin{aligned} pH &= \log_{10} 1/[H^+] \\ &= -\log_{10}[H^+] \end{aligned}$$

Expression of pH

Because pH is a logarithmic function, *do not* calculate or report as an average value. An "average pH" has no meaning. Usually the range of pH values is given and sometimes the median value.

Importance of pH

pH controls many chemical reactions, including coagulation, disinfection, water softening, corrosion, biochemical reactions, and ammonia removal. It also indicates to the design engineer what construction materials to use. A question often asked in water treatment is "where do I put the pH meter?"

SELECTED SUPPLEMENTARY READINGS

Clement, B. 1992. Computers Can Reduce Langelier Index Test Time. *Opflow,* 18(3):1.

Gordon, G., W.J. Cooper, R.G. Rice, and G.E. Pacey. 1992. *Disinfectant Residual Measurement Methods.* 2nd ed. Denver, Colo.: American Water Works Association and American Water Works Association Research Foundation.

Jensen, J.N., and J.D. Johnson. 1989. Specificity of the DPD and Amperometric Titration Methods for Free Available Chlorine. *Jour. AWWA,* 81(12):59.

Manual M12, Simplified Procedures for Water Examination. 2001. Denver, Colo.: American Water Works Association.

Manual of Instruction for Water Treatment Plant Operators. 1975. Albany, N.Y.: New York State Department of Health.

Manual of Water Utility Operations. 8th ed. 1988. Austin, Texas: Texas Water Utilities Association.

Methods of Chemical Analyses of Water and Wastes. 1984. EPA-600/4-79-020. Cincinnati, Ohio: US Environmental Protection Agency.

Standard Methods for the Examination of Water and Wastewater. 20th ed. 1998. A.D. Eaton, L.S. Clesceri, and A.E. Greenberg, eds. American Public Health Association, American Water Works Association, and Water Environment Federation.

Water Quality and Treatment. 5th ed. 1999. New York: McGraw-Hill and American Water Works Association (available from AWWA).

Organic Contaminants

All organic compounds or contaminants contain carbon in combination with one or more elements. A group of organic compounds called *hydrocarbons* contain only carbon and hydrogen (Figures 7–1 and 7–2). Many organics contain carbon, hydrogen, and oxygen. Naturally occurring organic compounds often contain low concentrations of nitrogen, phosphorus, and sulfur. Synthetic organic compounds may contain halogens, e.g., chlorine or fluorine, and inorganic metals.

NATURAL ORGANIC SUBSTANCES

Organic compounds differ from inorganic metallic and non-metallic compounds. Organic compounds

- are combustible,
- have lower melting and boiling points,
- are only slightly soluble in water,
- exhibit isomerism where more than one compound may exist for a chemical formula,
- have very high molecular weights,
- serve as substrate or food for bacteria, and
- have slower reaction rates.

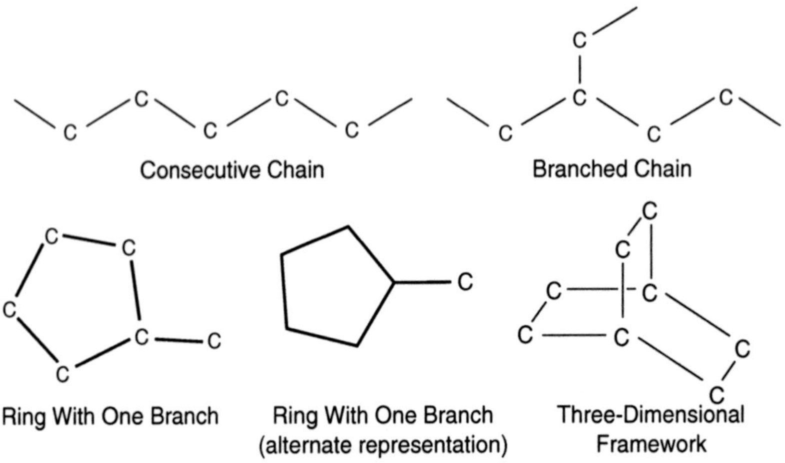

FIGURE 7-1 Typical arrangement of carbon atoms

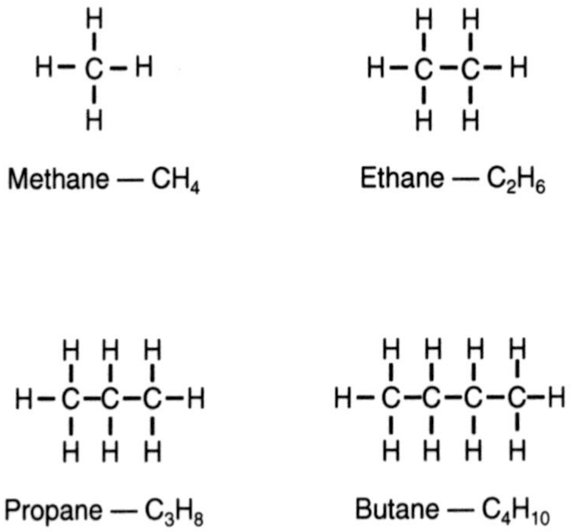

Consecutive Chain

Branched Chain

Ring With One Branch

Ring With One Branch
(alternate representation)

Three-Dimensional
Framework

Methane — CH_4

Ethane — C_2H_6

Propane — C_3H_8

Butane — C_4H_{10}

FIGURE 7-2 Typical hydrocarbons in chain configuration

Organic compounds find their way into water from three sources. The first source is humic materials from plants and algae, microorganisms and their secretions, and hydrocarbons. A few of the aromatic hydrocarbons may cause adverse health effects. Humic materials are precursors in the formation of trihalomethanes (THMs).

The second source is from domestic and commercial activities, wastewater treatment plant and industrial effluents into surface waters, such as rivers.

The third source is through reactions that occur during water treatment and transmission.

Groundwater

It is rather rare for groundwater sources to contain objectionable levels of natural organic compounds, but one situation in which such levels can occur is in a relatively shallow well overlain by a current or previously swampy area. If the taste, odor, and/or color is excessive, treatment may have to be provided to make the water palatable in the same manner as for a surface water source. It is not likely that such naturally occurring contaminants alone will create a serious health hazard. In such cases, if the amount of organic matter in the groundwater is relatively modest, it may not adversely affect taste, odor, or color. However, it might create excessive levels of disinfection by-products (DBPs) when the water is chlorinated.

Groundwater is also occasionally contaminated by naturally occurring hydrocarbons. In areas where natural gas and oil come in contact with aquifers, the water may be slightly contaminated but still useable with treatment, as a drinking water source. If the water is heavily contaminated, it will probably not be a suitable water source.

Surface Water

In general, surface waters are more prone to contamination by natural organic compounds than are groundwaters. The various types of vegetation growing in the watershed are one source of contamination. Many water systems regularly experience operational problems caused by decaying leaves and plants that have been washed off farms and forests during heavy rains in the spring and fall. This organic matter is generally decomposed by biological action and breaks down eventually into carbon dioxide and water. However, some organic compounds are quite complex and persist in the water environment for some time. For example, humic acid, derived from the decomposition of plant matter, is found in most surface waters and does not readily biodegrade (break down).

Microorganisms are another source of organic compounds in water. In addition to cellular matter, many plants and microorganisms release organic matter into a water source through their metabolic processes.

Various types of algae and vegetation flourishing in a lake or reservoir can also be the source of objectionable organic compounds in water. If the concentration of this vegetation is low, it usually has no adverse effect on drinking water quality. However, if there is a sudden die-off of the vegetation, water quality can become extremely bad. Health effects of large quantities of certain algae may include skin irritation, promotion of hay-fever-type allergies, and outbreaks of gastrointestinal illness.

Serious taste-and-odor problems can also be caused when a reservoir becomes stratified and matter near the bottom that has decomposed anaerobically (in the absence of free oxygen) is brought into the water system. Excessive amounts of algae in source water can also cause water treatment problems, such as taste and odor, filter clogging, and formation of slime in the treatment plant.

DBPs form when water containing organic substances is disinfected. In most cases, the organic substances are naturally occurring, such as humic and fulvic acids resulting from decaying vegetation. A group of chlorinated organic compounds called THMs was one of the first products of the reaction of chlorine with humic substances to be recognized. The principal THMs of concern are chloroform, bromodichloromethane, chlorodibromomethane, and bromoform. At one time, chloroform was widely used in cough medicine and other medications, but its use was discontinued when research determined that it is carcinogenic. The other THMs are also suspected of being carcinogens or have been demonstrated to have other adverse health effects. Thus, the various THMs are regulated as a group, with a maximum contaminant level (MCL) established for total THMs.

As more knowledge about DBPs develops, additional regulations limiting their concentration in finished water are expected. The next DBPs being considered for regulation are a group of five haloacetic acids, abbreviated as HAA5. DBPs are discussed in more detail in chapter 1.

Domestic and commercial activities contribute synthetic organic chemicals (SOCs) to wastewater discharges, agricultural runoff, urban runoff, and leachate from contaminated soils. Most of the organic contaminants identified in water supplies as having adverse health concerns are part of this group. They include pesticides (such as atrazine and aldicarb), solvents and metal degreasers (such as trichlorobenzene, tetrachloroethylene, trichloroethylene, and trichloroethane), and a family of compounds formerly in wide use, the polychlorinated biphenyls.

Organic contaminants formed during water disinfection include by-products such as THMs (e.g., chloroform) and HAAs (e.g., di- and trichloroacetic acids). Other compounds, such as acrylamide or epichlorohydrin, are components of coagulants (e.g., polyacrylamide) that can leach out during treatment. During finished water transmission, undesirable components of pipes, coatings, linings, and joint adhesives, such as polynuclear aromatic hydrocarbons (PAHs), epichlorohydrin, and solvents, have been shown to leach into water.

SYNTHETIC ORGANIC SUBSTANCES

The term synthetic organic chemicals has become a regulatory rather than a chemical description. It has evolved to distinguish a group of mostly volatile organic chemicals (VOCs), regulated first under the 1986 amendments of the federal Safe Drinking Water Act, from "SOCs" regulated under Phase 2 and later regulations. However, some of the SOCs are also VOCs (e.g., ethylbenzene, styrene, toluene, and xylenes, and the fumigant pesticides). The bulk of SOCs are pesticides but also include the PAHs, the polychlorinated biphenyls, and two water treatment polymers.

HEALTH EFFECTS OF ORGANIC CHEMICALS

The US Environmental Protection Agency (USEPA) has designated three health-effects categories for organic chemicals:

- Category I—It is known, or there is strong evidence, that the chemical is a carcinogen.

- Category II—There is limited but not positive evidence that the chemical is a carcinogen, and there are other known adverse health effects.

- Category III—There is no firm evidence that the chemical is a carcinogen, but there are other known adverse health effects.

Noncarcinogens

To the water system operator, the principal significance of a chemical's carcinogenic status is the way the maximum contaminant level goal (MCLG) is established. For noncarcinogens, the MCLG is a firm number, indicating the level of the contaminant that health-effects experts consider acceptable in drinking water. The MCL is then set at the same level as the MCLG, or as close to it as is considered technically achievable. The MCLG for these contaminants will be changed only if new information on their toxicity to humans becomes available.

The noncarcinogenic effects of organic chemicals on humans vary, but the principal ones are damage to the liver, kidneys, cardiovascular system, and central nervous system.

Carcinogens

For carcinogens, USEPA policy is that the MCLG must be zero. In other words, it is presently assumed that *any* exposure to the chemical could cause cancer, so ideally none of the chemical would be present in drinking water.

In the real world, though, there are two restrictions in controlling carcinogens: (1) the ability to detect the chemical by reasonable and reliable laboratory technique, and (2) the technology to remove the chemical from water if it is found to be present. These factors are considered when the MCLs are established, and the MCLs are set as close to the MCLGs as experts consider to be realistically achievable.

From time to time, then, USEPA must review all MCLs for carcinogens, and, if the factors considered in setting the MCL have changed, the MCL will be changed. In short, the intention is to continually edge the MCL for carcinogens closer to zero, so the allowable level is likely to be changed periodically.

MEASUREMENT OF ORGANIC COMPOUNDS

No single analytical method is capable of measuring all of the organic substances in a water sample. However, available analytical methods can be grouped into two categories, general and specific.

General Analytical Methods

Threshold odor tests, flavor profiles, and color determinations, described in chapter 5, have been used in the water utility industry for many years to obtain general measures of the levels of natural organic compounds in water. Two other methods used occasionally in monitoring water quality are ultraviolet light absorbance and fluorescence. These tests are used in some plants for control of organic compound removal processes because the measurement can be made quickly and easily.

Another test commonly used to determine the overall organic compound content of water is the measurement of total organic carbon (TOC). The typical concentration of TOC in water sources ranges from less than 0.5 mg/L to more than 10 mg/L. Highly colored water may have a TOC concentration of more than 30 mg/L.

Total organic halogen (TOX) is a relatively new measurement that is being used increasingly because it is specific to halogenated organic compounds. The presence of TOX in a sample is an almost certain indication of the presence of either synthetic organic compounds or DBPs. Only fairly sophisticated laboratories are presently capable of carrying out the procedures for TOC and TOX determinations.

Specific Analytical Methods

The list of organic compounds that have been identified in drinking water samples has grown from approximately 200 in 1975 to several thousand today, and it is constantly lengthening. In many cases, of course, a compound may have been identified only in isolated samples or at extremely low concentrations. However, the growth in the list is primarily because of steadily increasing improvements in analytical methods.

The three fundamental steps in gas chromatography are

1. Extraction and concentration of the organic compounds in the sample,
2. Separation of the extracted organic compounds, and
3. Detection of individual compounds.

Gas chromatography (Figure 7–3) requires very specialized equipment, detailed procedures, and trained operators, but in general the three steps are as follows.

Extraction and concentration

Organic substances are first extracted from a water sample. One method uses an organic solvent, such as methylene chloride or pentane. This process is called *liquid–liquid extraction*. Another method strips the organic compounds out of the sample using an inert gas, such as nitrogen or helium. This process is called gas purge.

Separation

The complex mixture must then be separated into its individual organic components. This process is carried out with a gas chromatograph or a high-performance liquid chromatograph. Chromatographs have a column of long, thin tubing through which individual organic compounds are driven off the sample as the temperature is slowly elevated. Thus, these processes may be viewed as sophisticated distillation or separation functions.

Detection

As the chromatograph separates the organic compounds by the temperature at which they are vaporized, they travel to a detector. Several types of detectors are available, each with certain advantages and disadvantages. The types in general use include flame ionization, electron capture, electrolytic conductivity, photoionization, and mass spectrometry. An organic compound is identified by comparing the signal the detector obtains (shown graphically in a gas chromatograph generated by the detector) with known standards for the compound. This process is generally aided by a computer connected to the equipment.

Figure 7–4 shows a chart produced by a GC, showing the presence of trihalo-methanes in a water sample.

Sampling for Organic Compounds

The location of sampling points in a water distribution system is very important; certain points should definitely not be used. Some typical locations to be avoided are described in the following list.

- Public restrooms should not be used as sampling locations because the de-odorizer commonly used in restrooms contains an organic chemical that may be in the air in sufficient concentration to contaminate the water sample.

- Gasoline service stations should be avoided because of the prevalence of petroleum products that could be in the air or that could have gotten on the sampling faucet.

- Any location where there are unusual odors, such as a freshly painted room, or where there is a smell from cleaning materials, should be avoided.

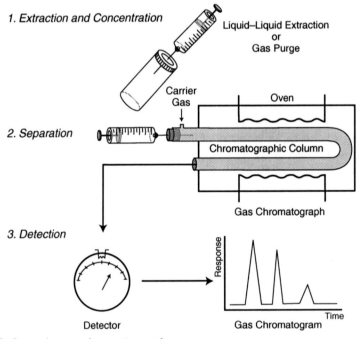

FIGURE 7-3 Steps in gas chromatography

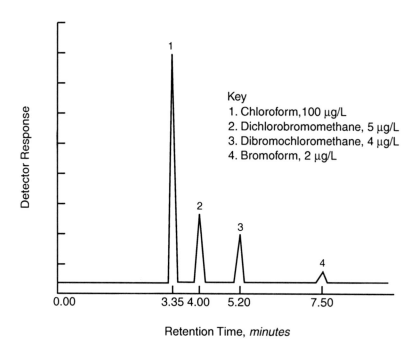

FIGURE 7–4 Sample readout from a gas chromatograph

- A location where a pump or piping has recently been installed or repaired should be avoided because of the possibility that organic solvents may have been used in the plumbing.

Ultraclean glass vials having lids with polytetrafluoroethylene (PTFE; trade name Teflon®) liners are used for collecting organic compound samples. The samples must be collected so that there is zero headspace in the vial; in other words, there must be no bubble of air in the vial after it is filled. If any air remains in the vial, a portion of the more volatile organic compounds will come out of solution and into the air space, which will cause inaccurate analysis of the water sample.

Each sample container must be completely labeled. A general rule is that the description of the sampling site must be complete enough so that a person unfamiliar with the initial sampling could return and collect a repeat sample from the same location if necessary.

SELECTED SUPPLEMENTARY READINGS

Manual of Water Utility Operations. 8th ed. 1988. Austin, Texas: Texas Water Utilities Association.

Standard Methods for the Examination of Water and Wastewater. 20th ed. 1998. A.D. Eaton, L.S. Clesceri, and A.E. Greenberg, eds. Washington, D.C.: American Public Health Association, American Water Works Association, and Water Environment Federation.

Water Quality and Treatment. 5th ed. 1999. New York: McGraw-Hill and American Water Works Association (available from AWWA).

8

Radiological Contaminants

Each of the forms of radiation reacts somewhat differently within the human body.

One of the more significant public health concerns regarding drinking water is the relatively high level of natural radioactivity found in some water sources. Most radioactivity in water is caused by natural causes, but there is also a threat of radionuclide contamination from various industrial and medical processes.

The harmful effects to a living organism of consuming water containing radioactivity are caused by the energy absorbed by the cells and tissues of the organism. This absorbed energy (or dose) produces chemical decomposition of the molecules present in the living cells. Each of the forms of radiation reacts somewhat differently within the human body.

RADIOACTIVE MATERIALS

A radioactive atom emits alpha particles, beta particles, and gamma rays and is shown in Figure 8–1.

Alpha Particles (Radiation)

Alpha particles are the most prevalent naturally occurring radionuclide present in drinking water and are therefore of the greatest concern. Alpha (α) particles are the heaviest particles.

Alpha radiation is not true electromagnetic radiation like light and X rays. It consists of particles of matter. Alpha particles are doubly charged ions of helium. Although they are propelled from the nucleus of atoms at approximately 10 percent of the speed of light, they do not travel much more than

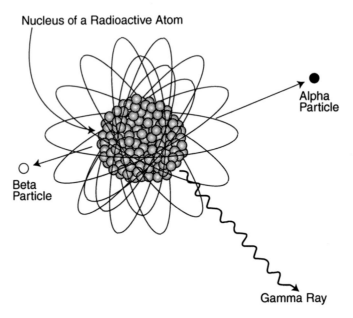

Nucleus of a Radioactive Atom

Alpha
Particle

Beta
Particle

Gamma Ray

TABLE 8–1 Emissions from the nucleus of a radioactive atom

10 cm in air at room temperature. They are stopped by an ordinary sheet of paper. The alpha particles emitted by a particular element are all released at the same velocity. The velocity varies, however, from element to element. The alpha particles have extremely high ionizing action within their range.

Beta Radiation
Beta radiation consists of negatively charged particles—electrons—that move at speeds ranging from 30 to 99 percent of the speed of light. The penetrating power of beta radiation depends on its speed. It can travel several hundred feet in air and can be stopped with a few millimeters of aluminum. The ionizing power of beta radiation is much less than that of alpha radiation.

Gamma Radiation
Gamma radiation is true electromagnetic radiation which travels with the speed of light. It is similar to X rays but has a shorter wavelength and therefore greater penetrating power, which increases as the wavelength decreases. Proper shielding

from gamma radiation requires several centimeters of lead or several feet of concrete. The unit of gamma radiation is the *photon*.

Units of Radioactivity

The unit of radioactivity is the *curie*. Formerly, it was considered to be the number of disintegrations occurring per second in one gram of pure radium. Because the constants for radium are subject to revision from time to time, the International Radium Standard Commission has recommended the use of a fixed value, 3.7×10^{10} disintegrations per second, as the *standard curie* (Ci).

The curie is used mainly to define quantities of radioactive materials. A curie of an alpha emitter is that quantity which releases 3.7×10^{10} alpha particles per second. A curie of a beta emitter is that quantity of material which releases 3.7×10^{10} beta particles per second, and a curie of a gamma emitter is that quantity of material which releases 3.7×10^{10} photons per second. The curie represents such a large number of disintegrations per second that the millicurie (mCi), microcurie (μCi), and picocurie (pCi), corresponding to 10^{-3}, 10^{-6}, 10^{-9}, and 10^{-12} curie, respectively, are more commonly used.

The roentgen is a unit of gamma or X-ray radiation intensity. It is of value in the study of the biological effects of radiation that result from ionization induced within cells by the radiations. The roentgen is defined as the amount of gamma or X-ray radiation that will produce in one cubic centimeter of dry air, at 0°C and 760 mm pressure, one electrostatic unit (esu) of electricity. This is equivalent to 1.61×10^{12} ions pairs per gram of air and corresponds to the absorption of 83.8 ergs of energy.

The roentgen is a unit of the total quantity of ionization produced by gamma or X rays, and dosage rates for these radiations are expressed in terms of roentgens per unit time.

With the advent of atomic energy involving exposure to neutrons, protons, and alpha and beta particles that also have effects on living tissue, it has become necessary to have other means of expressing ionization produced in cells. Three methods of expression have been used.

The *roentgen equivalent physical (rep)* is defined as that quantity of radiation (other than X-rays or radiation) which produces in one gram of human tissue ionization equivalent to the quantity produced in air by one roentgen of radiation or X rays (equivalent to 83.8 ergs of energy). The rep has been replaced largely by the term rad, which has wider application.

The *radiation absorption dose (rad)* is a unit of radiation corresponding to an energy absorption of 100 ergs per gram of any medium. It can be applied to any type and energy of radiation that leads to the production of ionization. Studies on the radiation of biological materials have shown that the roentgen is approximately equivalent to 100 ergs/g of tissue and can be equivalent to 90 to 150 ergs/g of tissue, depending on the energy of the X and radiation and type of tissue. The rad, therefore, is more closely related to the roentgen than is the rep, in terms of radiation effects on living tissues, and is the term preferred by biologists.

The rad represents such a tremendous radiation dosage, in terms of permissible amounts for human beings, that another unit has been developed specifically for man. The term *roentgen equivalent man (rem)* is used. It corresponds to the amount of radiation that will produce an energy dissipation in the human body that is biologically equivalent to one roentgen of radiation of X rays, or approximately 100 ergs/g. The recommended maximum permissible dose for radiation workers is 5 rem/year and for nonradiation workers ½ rem/year.

RADIOACTIVE CONTAMINANTS IN WATER

Humans receive a radiation dose of about 200 mrem from all sources each year, and the US Environmental Protection Agency (USEPA) estimates that, on the average, as much as 3 percent of this dose comes from drinking water. Local conditions can, of course, greatly alter this proportion.

The radioactive substances that are of concern as drinking water contaminants are

- radium
- uranium
- radon, and
- artificial radionuclides.

Radium

Radium is the most common radionuclide of concern in drinking water. Naturally occurring radium is leached into groundwater from rock formations, so it is present in water sources in those parts of the country where there is radium-bearing rock. It may also be found in surface water as a result of runoff from mining and industrial operations where radium is present in the soil. The two isotopes (variations) of radium of concern in drinking water are radium 226, which emits principally alpha particles, and radium 228, which emits beta particles and alpha particles from its daughter decay products.

Uranium

Naturally occurring uranium is found in some groundwater supplies as a result of leaching from uranium-bearing sandstone, shale, and other rock. Uranium may also occasionally be present in surface water, carried there in runoff from areas with mining operations. Uranium may be present in a variety of complex ionic forms, depending on the pH of the water.

Radon

Radon is a naturally occurring radioactive gas. It cannot be seen, smelled, or tasted. Radon comes from the natural breakdown (radioactive decay) of uranium. It is the direct radioactive-decay daughter of radium 226. The highest concentrations of radon are found in soil and rock containing uranium. Significant concentrations, from a health standpoint, may be found in groundwater from any type of geologic formation, including unconsolidated formations.

Outdoors, radon emitted from the soil is diluted to such low concentrations that it is not of concern. However, when it is liberated inside a confined space, such as a home or office building, radon can accumulate to relatively high levels, and inhalation of the gas is considered a health danger. Most cases of excessive levels of radon in buildings are caused by the gas seeping through cracks in concrete floors and walls. In areas where high levels of radon in the soil are a problem, foundation ventilation should be installed to reduce the concentration of radon entering buildings.

The problem from a public water supply standpoint is that, if radon is present in the water, a significant amount of the gas will be liberated into a building as water is used. Showers, washing machines, and dishwashers are particularly efficient in transferring radon gas into the air. The radon released from the water adds to the radon that seeps into a building from the soil, adding to the health danger.

Artificial Radionuclides

Significant artificial radionuclide levels have been recorded in surface waters as a result of atmospheric fallout following nuclear testing, leaks, and disasters. Otherwise, surface water generally contains little or no radioactivity. Potential sources of serious water contamination are accidental discharges from facilities using radioactive materials, such as power stations, industrial plants, waste-disposal sites, or medical facilities. State and federal nuclear regulatory agencies monitor all uses of radioactive materials to prevent such discharges. If an accidental discharge of artificial radionuclides takes place, the elements most likely to be present are strontium 90 and tritium.

ADVERSE HEALTH EFFECTS OF RADIOACTIVITY

The effects of excessive levels of radioactivity on the human body include developmental problems, nonhereditary birth defects, genetic effects that might be inherited by future generations, and various types of cancer. All radionuclides are considered to be carcinogens (cancer-causing agents).

Radium is chemically similar to calcium, so about 90 percent of ingested naturally occurring radium goes to the bones. Consequently, the primary risk from radium ingestion is bone cancer. Uranium has not definitely been proven to be carcinogenic, but it accumulates in the bones, much as radium 228 does. The principal adverse effect of uranium has been shown to be toxicity to human kidneys.

Inhaled radon is considered to be a cause of lung cancer. Radon is also thought to have some noncarcinogenic effects on internal body organs when ingested.

Although the proportion of radon added to a building by the water supply is usually relatively small in comparison to the amount that seeps into the building from the soil, radon in drinking water is still significant because of the large number of people who are being exposed. USEPA estimates that between 1 and 5 million homes in the United States may have significantly high levels of radon contamination and that between 5,000 and 20,000 lung cancer deaths a year may be attributed to all sources of radon.

RADIONUCLIDE MONITORING REQUIREMENTS

The level of restrictions that should be placed on radioactivity in drinking water has been the subject of extensive research and much debate. Some experts feel that the requirements should be much more restrictive, and others are of the opinion that the danger is not serious and the requirements should be relaxed.

Another factor that has contributed to the dilemma of regulation is the high cost of radionuclide analyses. Although some progress has been made in the form of improved equipment and automated operation, analyses still require expensive equipment and trained staff to operate it. The cost of analyses is kept as low as possible by the requirement of an initial scan to determine if significant radioactivity is present. Only if the level is higher than a specified amount is it necessary to progress to further analyses.

Interim Regulations

The maximum contaminant levels (MCLs) and monitoring requirements for radionuclides that were promulgated by USEPA as part of the National Interim Primary Drinking Water Regulations in 1976 were basically as follows:

- Combined radium 226 and radium 228—5 pCi/L.
- Gross alpha particle activity (including radium 226 but excluding radon and uranium)—15 pCi/L.

- Average annual concentration of beta particle and photon radioactivity from manufactured radionuclides—to produce an annual dose equivalent to no greater than 4 mrem/year.

Gross alpha particle activity is used as an initial scan, and if it is less than 3 pCi/L, the state may allow a system to reduce monitoring frequency in the future. If it is more than 3 pCi/L, additional analyses for specific radionuclides are required.

Proposed Regulation Changes

In 1991, USEPA proposed a revised regulation for radionuclides in drinking water. The proposal suggests some modified as well as some new MCLs for radon and uranium. The proposed new standards are shown in Table 8–1. It is emphasized that these standards were *proposed* but still not promulgated as of the publication of this volume. The radon MCL has been particularly debated. It is estimated that at least 32,000 community and nontransient, noncommunity water systems in the United States will be out of compliance with the radon standard if it is established at the proposed level of 300 pCi/L.

TABLE 8–1 USEPA's proposed revised regulation for radionuclides in drinking water

Contaminant	MCL	Notes
Radium 226	20 pCi/L	This has been raised from the interim standard of 5 pCi combined radium 226 and 228
Radium 228	20 pCi/L	This has been raised from the interim standard of 5 pCi combined radium 226 and 228
Uranium	20 µg/L	New standard; based on chemical toxicity, so is a weight-based standard
Radon	300 pCi/L	New standard (modified from 200 pCi/L in 1994)
Alpha emitters	15 pCi/L	Called adjusted gross alpha; calculated as gross alpha activity, minus radium 226 and uranium activity
Beta particle	4 mrem/year	Primarily applicable to manufactured radiation and photon emitters

SELECTED SUPPLEMENTARY READINGS

Lowry, J.D. 1991. Measuring Low Radon Levels in Drinking Water Supplies. *Jour. AWWA*, 83(3):149.

Pontius, F.W. 1992. USEPA's Proposed Radon MCL: Too High, Too Low, or Just Right? *Jour. AWWA, 84*(10):20.

Pontius, F.W. 1994. Disposal of Radioactive Residuals Requires Careful Planning. *Jour. AWWA,* 86(11):18.

Radioactivity in Drinking Water. 1991. Criteria and Standards Division, US Environmental Protection Agency. Washington, D.C.: USEPA.

Water Quality and Treatment. 5th ed. 1999. New York: McGraw-Hill and American Water Works Association (available from AWWA).

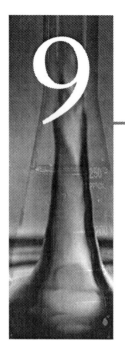

Customer Complaint Investigation

Responding positively to the concerns of customers regarding water quality can be very beneficial for water utilities in many ways. The process of responding gives operators an additional tool in tracking water quality through the distribution system. A customer complaint may be an early indication of a quality problem. In addition, customers' confidence in the quality of the water may be strengthened by professionally conducted investigations of their concerns.

A visit from a well-informed utility employee can certainly improve public relations. Such visits are also opportunities to educate customers regarding the utility's operations and water quality.

The problem might be real or only perceived; regardless, the customer has a problem he would like solved.

GENERAL PRINCIPLES

An investigation should always be undertaken with the philosophy that the customer would not be calling if there was not a problem. The problem might be real or only perceived; regardless, the customer has a problem he would like solved. The caller may be angry, frustrated, embarrassed, or uncomfortable about calling, so the receiver of the call should allow the customer to explain the reason for calling before starting to ask questions. If the customer is in a highly emotional state, this approach sometimes may give him/her an opportunity to vent his/her feelings so that he/she may then discuss the subject calmly.

The first order of business is to define the problem. The customer may not know how to explain exactly what is bothering her or exactly what questions to ask. The receiver should repeat to the customer what he heard: "You said, Mrs. Smith, that the water coming from your water faucets is brown and has an odor" or "You say that your family is ill from drinking the water?" With the problem established and agreed upon, further questions may be asked to gain details.

Complaint Form
It is helpful for the receiver to have a form to fill out while taking the complaint. The form should have three parts:

1. The receiving information.
2. The investigation results, including lab results.
3. A description of the final disposition, including the customer's satisfaction with the investigation.

Investigation
Although it is often difficult to do, the investigator should approach the problem with an open mind, with no preconceived notions about any part of the investigation. The customer should be asked again to explain the problem, and the investigation should be limited to that problem only.

In most investigations, water samples should be collected at a cold-water tap, either to confirm the problem or to prove that the water being delivered to the premises matches the general condition of the water in the distribution system. Temperature and chlorine residual should be tested onsite, and a general chemical sample (hardness, pH, alkalinity, and the like) and a bacteriological sample should be collected for analysis in the lab.

If the solution to the problem is obvious, the customer should be informed immediately. If the solution is the customer's responsibility, the investigator should advise the customer about ways and means to implement the solution. If the solution is the utility's responsibility, the investigator should advise the customer how the utility will deal with the problem. If the problem is a perceived one—that is, not really a problem—the investigator must communicate this information tactfully to the customer.

Final Disposition
Regardless of the details of the investigation, the investigator should carry it through to a resolution. The customer should be notified of any laboratory results, kept advised of the investigation's progress, and contacted upon its conclusion to

ascertain his or her satisfaction. The final result should be a satisfied customer. The completed complaint form should be kept on file for future reference.

SPECIFIC COMPLAINTS

A vast majority of complaints fall into one or more of the following categories:

- Objectionable taste and/or odor
- Objectionable appearance of the water
- Stained laundry and plumbing fixtures
- Illness alleged to be caused by the drinking water

Taste and Odor

Surveys have shown that taste-and-odor complaints are the type received most frequently by most water utilities, especially utilities treating surface waters. Sources of taste- and odor-causing compounds may either be natural or caused by pollution. Natural taste- and odor-causing compounds result from plant growth or animal activities in the watershed or source water. Most natural taste- and odor-causing compounds produce fishy, earthy, or manure-type tastes and odors. Industrial and agricultural discharges into water sources generally produce chemical or medicinal tastes and odors.

Human perceptions of tastes and odors are highly variable. How individuals define what they taste and smell depends on many factors, such as the person's age, health, previous experiences, and other senses as they interact with taste and smell. An individual cannot describe an odor as "potato bin" if he/she has not smelled a potato bin. And it is not unusual for an individual to think he/she detects an odor in cloudy water because of what he/she sees. All these factors make the investigation of taste-and-odor complaints very tricky.

Receiving information

Once it has been established that taste and odor are the subjects of a complaint, the customer should be asked to describe what she/he tastes or smells. She/he should also be asked how long she/he has experienced the problem. This information is important for the investigator to know.

Investigating

The investigator should try tasting and/or smelling the water from the same faucets the customer has been using. If no taste or odor is detected in the cold water, water from the hot-water system can be tested. Many taste- and odor-causing compounds are volatile and can be tasted or smelled more readily from hot water.

It may also be possible that the taste or odor exists only in the hot water, in which case the problem can be immediately located in the residence's hot-water system.

If a customer detects an odor when it is already known that the water system is experiencing a taste-and-odor episode, the source of the problem can be explained to the customer over the phone. If a customer's detection of an odor appears to be an isolated case, further investigation is required. If the investigation reveals no odor, more investigation may still be necessary to convince the customer that the problem is only a perceived one. In any case, samples should be collected for study at the plant or laboratory.

In conducting the investigation, the investigator should attempt to imagine the many potential sources of taste and odor. Following is a list of some of the more probable causes and situations.

- A general taste-and-odor incident is occurring in the source water, and the caller is the first customer who has complained.
- The customer's water service is connected to a dead end or to a low-flow main in the distribution system, and stagnant water is being drawn into the residence's water service.
- A cross connection has drawn some foreign substance into the water system.
- Water system maintenance in the vicinity has stirred up stagnant water or sediment in the mains.
- Waste plumbing in the residence is what is actually causing the odor problem.
- The taste or odor is originating in the hot-water system in the residence.
- If a home water conditioner (water softener, carbon filter, or the like) is being used, it could be causing the problem.
- Customers who are in poor health or are elderly may be more inclined to imagine a problem.
- Customers may actually be tasting or smelling something, such as medication they are taking, that is not in the water supply.

Disposition of the complaint

After the problem has been identified, appropriate action must be taken. If the problem is within the jurisdiction of the utility, corrective measures must begin as soon as possible. If the problem is isolated within a residence, the investigator must work with the customer by advising her/him of the steps she/he can take to eliminate the problem. In all cases the customer must be kept advised of all the steps that are being taken to solve the problem, including the results of any laboratory testing. After the problem has been solved, the customer should be contacted to verify her/his satisfaction with the situation.

Physical Appearance

Customers generally expect clear, cold, odorless water to be available from their taps at all times. When the water deviates from this norm, they become concerned and will report their concern to the water utility. The physical appearance of water can be affected adversely by such things as excess air in the water, sediment from disturbed water lines, rust, particulate matter, bugs, or worms.

Receiving information

The receiver of the complaint call needs to obtain an exact description of the offending appearance of the water and when it was first noticed by the customer. On some calls, the receiver may be able to diagnose the problem and offer assistance over the phone, particularly if the cause is already known from a previous investigation in the same area.

Investigating

The investigator must observe the offending appearance of the water in order to develop information from which to draw conclusions and offer solutions. In some cases, material in the water may have to be analyzed chemically or under a microscope for proper identification. Once the material and the source have been identified, the investigator can take steps to correct the problem.

One rather common complaint received by some water systems is that the water is "cloudy" when all it contains is entrained air. Sometimes the callers are new customers who had not seen entrained air in the water where they lived before, but complaints may also come from old customers who may just be noticing the phenomenon for the first time. The problem occurs at certain times of the year when there is an excess of air entrained in the water. When a glass of water is filled, bubbles are released, making the water look cloudy when it is fresh from the tap. The cloudiness quickly clears, though, starting almost immediately at the bottom of the glass and moving upward until the water is completely clear in a minute or so.

Disposition of the complaint

Most complaints of dirty water due to suspended matter in the mains can be cleared up by flushing the distribution system and the customer's house plumbing. Regardless, once the problem and source have been identified, the investigator must follow through to a conclusion. Again, the utility must take action if the problem falls under its jurisdiction, and the investigator should suggest solutions to the customer if the problem is isolated in the residence.

Staining of Laundry and Plumbing Fixtures

Staining of laundry and plumbing fixtures can occur when the water contains iron, manganese, or copper in solution. It is relatively common for there to be some dissolved iron and manganese in groundwater. Also, under some relatively rare conditions there can be objectionable levels of iron in surface water.

When a groundwater system pumps directly from wells to the distribution system, the water will generally be clear as it comes from the customer's tap. However, after the water is exposed to air in a bathtub, toilet, or washing machine, iron will oxidize to red-brown ferric hydroxide precipitate. In some situations iron will be partially or completely oxidized in the water mains, and customers will get discolored water either continuously or sporadically.

The iron precipitate will cause laundered white clothes to have an off-white color, and brown stains will build up on porcelain fixtures. A particularly exasperating problem for customers is that, as they repeatedly scour the porcelain fixtures to remove the discoloration, they slowly break down the porcelain's surface glaze, exposing the more porous ceramic below; these areas then become discolored even faster and are harder to clean. When this has happened there is generally no recourse but to replace the fixtures.

Another common complaint from customers when the iron content is high is that coffee and tea turn so dark they look like ink. This darkness is caused by a reaction of the iron with the tannic acid in the beverages.

Manganese is often present with iron in groundwater and may cause similar staining problems, except that a dark-brown to black staining precipitate is formed.

Copper staining is usually most objectionable when it creates blue-green stains on plumbing fixtures. Copper staining is caused by aggressive water that dissolves copper from the customer's piping system.

Receiving information

The receiver of the complaint call needs to obtain a description of the problem and the customer's location. If the cause is already known from previous complaints, it may be possible to give the customer some advice over the phone on removing the stains or avoiding stains in the future.

Investigating

If the complaint is new for the system or for a particular area of the system, the investigator should visit the customer and observe the problem. In some cases, staining can occur as a result of local problems, such as a dead-end main, and special corrective action may be possible.

Disposition of the complaint

If the problem is only a local condition, it may be possible to correct it by flushing mains in the area. If the problem recurs regularly in the area, it may be necessary to set up a regular schedule for flushing the mains.

If the problem is found to be general throughout the system, the utility should take steps to provide treatment to prevent staining from occurring. Methods of iron and manganese control are covered in *Water Treatment*, another book in this series.

Illness Caused by Water

Illnesses can be attributed to drinking water, and customers have generally been made aware of this fact by their own education or by information obtained through the media or their doctor. Some customers will call the water utility after visiting a doctor and being told that one source of their illness could be drinking water.

Receiving information

Calls concerning waterborne illness may be some of the most difficult to handle. In many cases the customer is not sure of terminology and does not know what questions to ask to initiate the investigation. The receiver of the complaint must be very sensitive in attempting to gain information. Very seldom, with the exception of *Giardia*, will the infectious agent be known.

The chances of illness being caused by contamination of a well-run public water system are quite remote, but it does happen, so the customer's complaint cannot be immediately discounted. The receiver needs to determine the symptoms of the illness, how many people have been affected, and if the illness has been diagnosed by a physician. The receiver must be very careful not to sound as if he/she has medical knowledge as he responds to or asks questions.

Investigating

Generally, in customer-initiated calls, the customer is seeking to check on the drinking water as a potential source of his/her illness. The investigator's job is to provide the customer with enough information to allow him/her to reach his/her own conclusions regarding the water quality.

Even if the person has an illness that is known as a waterborne disease, such as giardiasis, the illness could have been contracted through a source other than the water system. The vast majority of cases are actually contracted through person-to-person contact, although an occasional case of giardiasis can be traced to a person's contact with untreated water during a camping or fishing trip.

Nevertheless, a sample for bacteriological analysis should be drawn from a cold-water tap along with a sample for general chemical analysis to set the customer's mind at ease. A chlorine residual test should be conducted in the presence of the customer and the results explained. The investigator should tell the customer that the bacteriological analysis will determine the presence or absence of coliform bacteria and that coliforms are an indicator for pathogenic microorganisms. It should further be explained that the chemical analysis will compare the customer's tap water with the water being served to her/him from the distribution system.

The investigator should also explain to the customer about how the water is treated and what precautions are taken to protect the water supply. The customer should be encouraged to consult with a physician if that has not been done.

If investigations are indicating a possibility that a waterborne illness is occurring, it is prudent and necessary that the utility notify the primacy agency and possibly also the state and local health departments and request their assistance.

Disposition of the Complaint

The results of the bacteriological and chemical analyses should be relayed to the customer as soon as possible and a discussion held as to her/his perception of the investigation. The complaint form should be filed for future reference.

SELECTED SUPPLEMENTARY READINGS

Hack, D.J. 1990. Phew! My Hot Water Smells Like Rotten Eggs. *Opflow,* 16(7):1.

Lay, T. 1989. Fight Consumer Fear, Suspicion with Straight Talk. *Opflow,* 15(8):3.

Reinert, R.H. 1992. Quality Is Defined by the Customer. *Jour. AWWA,* 84(8):20.

Glossary

AA See *atomic absorption spectrophotometer.*

absence See *negative sample.*

acidity Base neutralizing power.

agar A nutrient preparation used to grow bacterial colonies in the laboratory. Agar is poured into petri dishes to form agar plates or into culture tubes to form agar slants.

air-strip To remove gasses from water by passing large volumes of air through the water.

algae Primitive plants (one- or many-celled) that usually live in water and are capable of obtaining their food by photosynthesis.

alkalinity Acid neutralizing power.

alpha particle A positively charged particle given off by certain radioactive substances. It consists of two protons and two neutrons and is converted into an atom of helium by the acquisition of two electrons.

alum The most common chemical used for coagulation. It is also called aluminum sulfate.

anaerobic Characterized by the absence of air or free oxygen.

analytical balance A sensitive balance used to make precise weight measurements.

aspirate To remove a fluid from a container by suction.

aspirator A T-shaped plumbing fixture connected to a water faucet. It creates a partial vacuum for filtering operations.

atom The basic structural unit of matter; the smallest particle of an element that can combine chemically with similar particle of the same or other elements to form molecules of a compound.

atomic absorption spectrophotometer (AA) A spectrophotometer used to determine the concentration of metals in water and other types of samples.

atomic absorption spectrophotometric method An analytical technique used to identify the constituents of a sample by detecting which frequencies of light the sample absorbs.

autoclave A device used for sterilizing laboratory equipment by using pressurized steam.

autoclaved Sterilized with steam at elevated temperature and pressure.

bacterial aftergrowth Growth of bacteria in treated water after the water reaches the distribution system.

balance An instrument used to measure weight.

BAT See *best available technology.*

beaker A container with an open top, vertical sides, and a pouring lip used for mixing chemicals.

beam balance See *Single-pan balance.*

best available technology (BAT) The best technology, treatment techniques, or other means that are available for treatment of a water quality problem and that have been found to be practical under field conditions.

beta particle An electron ejected form the nucleus of certain radioactive substances.

biochemical oxygen demand (BOD) A measurement of the amount of oxygen used in the biochemical oxidation of organic matter over a specified time (usually 5 days) and at a specific temperature (usually 35°C). Used to indicate the level of contamination in water of contamination potential of a waste.

BOD See *biochemical oxygen demand.*

borosilicate glass A type of heat-resistant glass used for labware.

breakpoint The point at which the chlorine dosage has satisfied the chlorine demand.

breakpoint chlorination The addition of chlorine to water until the chlorine demand has been satisfied and free chlorine residual is available for disinfection.

buffering capacity The capability of water of chemical solution to resist a change in pH.

burette   A graduated glass tube fitted with a stopcock, used to dispense solutions during titration.

burner   A high-temperature-heating device that uses natural or bottled gas. Also called a *gas burner* or *Bunsen burner.*

calcium carbonate   Scale-forming substance in water.

calibrate   To adjust a measuring instrument so that it gives the correct result with a known concentration or sample.

carcinogen   Any substance that causes cancer.

chlorination   The process of adding chlorine to water to kill disease-causing organisms or to act as an oxidizing agent.

chlorine demand   The quantity of chlorine consumed by reaction with substances in water.

Ci   See *curie.*

coagulation   The water treatment process that causes very small suspended particles to attract one another and form larger particles. This process is accomplished by adding a chemical, called a coagulant, that neutralizes the electrostatic charges on the particles that cause them to repel one another.

coliform bacteria   A group of bacteria predominantly inhabiting the intestines of humans or animals but also occasionally found elsewhere. Presence of the bacteria in water is used as an indication of fecal contamination (contamination by human or animal wastes).

coliforms (total coliforms)   See *coliform bacteria.*

colony counter   An instrument used to count bacterial colonies for the standard plate count test.

color   A physical characteristic of water. Color is most commonly tan or brown due to oxidized iron, but contaminants may cause other colors, such as green or blue. Different from turbidity, which is the cloudiness of water.

color comparator   A device used for tests such as chlorine residual or pH. Concentrations of constituents are determined by visual comparison of a permanent standard (usually sealed in glass or plastic) and a water sample.

colorimeter   An instrument that measures the concentration of a constituent in a sample by measuring the intensity of color in that sample. The color is usually created by mixing a chemical reagent with the water sample according to a specific test procedure.

colorimetric method   Any analytical method that measures a constituent in water by determining the intensity of color in the water. The color is usually produced when a chemical solution specified by the particular procedure is added to the water.

color unit (cu)   The unit of measure of the color of water, measured by comparing the color of a water sample with the color of a standard solution.

combined chlorine residual   The chlorine residual produced by the reaction of chlorine with substances in the water. Because the chlorine is "combined" it is not as effective a disinfectant as free chlorine residual.

community public water system   As defined by the National Primary Drinking Water Regulations, a system that serves at least 15 service connections or at least 25 full-time residents 60 or more days per year.

complaint investigation   A professionally conducted investigation of a customer's water quality complaint.

completed test   The third major step of the multiple-tube fermentation method. This test confirms that positive results from the presumptive test are due to coliform bacteria. See also *confirmed test; presumptive test.*

composite sample   A series of individual or grab samples taken at different times from the same sampling point and mixed together.

compound microscope   A microscope with two or more lenses.

confirmed test   The second major step of the multiple-tube fermentation method. This test confirms that positive results from the presumptive test are due to coliform bacteria. See also *completed test; presumptive test.*

cross-connection   Any connection between a safe drinking water supply and a nonpotable water or other fluid. Also called cross contamination.

$C \times T$ value   The product of the residual disinfectant concentration, C, in milligrams per liter, and the corresponding disinfectant contact time, T, in minutes, or $C \times T$. Minimum $C \times T$ values are specified by the Surface Water Treatment Rule as a means of enduring adequate kill or inactivation of pathogenic microorganisms in water.

cu   See *color unit.*

culture tube   A hollow, slender glass tube with an open top and a rounded bottom used in microbiological testing procedures such as the multiple-tube fermentation test.

curie (Ci)   The activity of 1 g of radium, or 3.7×10^{10} disintegrations/sec.

cyst   A resistant form of a living organism.

D/DBPs   See *disinfectants–disinfection by-products.*

deionizer   A device used to remove all dissolved inorganic ions from water.

deluge shower   A safety device used to wash chemicals off the body quickly.

desiccator   A tightly sealed container used to cool heated items before they are weighed. This procedure prevents the items from picking up moisture in the air and increasing the weight.

dilution bottle   A type of heat-resistant glass bottle used for diluting bacteriological samples before analysis. Also called *milk dilution bottle* or *French square.*

disinfectants/disinfection by-products (D/DBPs)   A term used in connection with state and federal regulations designed to protect public health by limiting the concentration of either disinfectants or the by-products formed by the reaction of disinfectants with other substances in the water (such as trihalomethanes).

disposition of complaint   An official completion of a complaint investigation, including an assessment of customer satisfaction.

dissolved oxygen (DO)   The oxygen dissolved in water, wastewater, or other liquid, usually expressed in milligrams per liter, parts per million, or percent of saturation.

dissolved solids   Any material that is dissolved in water and can be recovered by evaporation of the water after filtering the suspended material. Also called *filterable residue.*

DO   See *dissolved oxygen.*

double-pan balance   A balance that weighs material by counterbalancing material placed on one pan with brass weights placed on the other pan.

EC   See *electrical conductivity.*

E. coli   See *Escherichia coli.*

EDTA (ethylenediaminetetraacetic acid)   A chemical used to sequester, or tie up, calcium and magnesium ions, used in the hardness test.

electrical conductivity (EC)   A test that measures the ability of water to transmit electricity. Electrical conductivity is an indicator of dissolved solids concentration. Normally an EC of 1,000 mho/cm^2 indicates a dissolved solids concentration of 600–700 mg/L.

electrode method   Any analytical procedure that uses an electrode connected to a millivoltmeter to measure the concentration of constituent in water.

electrophotometer   A photometer that uses different colored-glass filters to produce wavelengths desired for analyses. Also called a *filter photometer.*

Enhanced Surface Water Treatment Rule (ESWTR)   A revision of the original Surface Water Treatment Rule that includes new technology and requirements to deal with newly identified problems.

equilibrium   A balanced condition in which the rate of formation and the rate of consumption of a constituent of constituents are equal.

Erlenmeyer flask   A bell-shaped container used for heating and mixing chemicals and culture media.

Escherichia coli (E. coli)   A bacteria of the coliform group used as a substitute for fecal coliforms in the regulations of the Total Coliform Rule.

ESWTR   See *Enhanced Surface Water Treatment Rule.*

evaporating dish   A glass or porcelain dish in which samples are evaporated to dryness using high heat.

eyewash   A safety device used to wash chemicals from the eyes. One type of device resembles a drinking fountain and directs a gentle spray of water into each eye.

fecal coliform   A bacteria of the coliform group indicative of fecal contamination. The presence of fecal coliform in a water sample is a reportable violation of the Total Coliform Rule.

filter (laboratory)   A porous layer of paper, glass fiber, or cellulose acetate used to remove particulate matter from water samples and other chemical solutions.

filter paper   Paper with pore size usually between 5 and 10 μm used to clarify chemical solutions, collect particulate matter, and separate solids from liquids.

filter photometer   See *electrophotometer.*

filterable residue   See *dissolved solids.*

filtering crucible   A small porcelain container with holes in the bottom, used in the total suspended solids test. Also known as a Gooch crucible.

flaming   The process of passing a flame over the end of a faucet in order to kill bacteria before taking a water sample for bacteriological sampling. The procedure is no longer recommended because it may damage the faucet and is of questionable benefit.

flask   A container, often narrow at the top, used for holding liquids. There are many types of flasks, each with its own specific name and use.

flocculation   The water treatment process following coagulation, which uses gentle stirring to bring suspended particles together so they will form larger, more settleable clumps called floc.

flow-proportional composite   A composite sample in which individual sample volumes are proportional to the flow rate at the time of sampling.

free available chlorine residual   The residual formed once all the chlorine demand has been satisfied. The chlorine no longer combines with other constituents in the water and is "free" to kill microorganisms.

French square   See *dilution bottle.*

full-face shield   A shatterproof plastic shield worn to protect the face from flying particles and chemicals.

fume hood   A large enclosed cabinet equipped with a fan to vent fumes from the laboratory. Mixing and heating of chemicals are done under the hood to prevent fumes from spreading through the laboratory.

funnel   A utensil used in the laboratory for pouring liquids into flasks and other containers. Laboratory funnels are either glass or plastic.

gamma ray   A form of electromagnetic radiation emitted in nuclear decay.

gas chromatography (GC)   A technique used to measure the concentration of organic compounds in water.

gas chromatograpy–mass spectrophotometry (GC–MS)   A very sophisticated analytical technique for analyzing and identifying organic compounds.

GC   See *gas chromatography.*

GC–MS   See *gas chromatography–mass spectrophotometry.*

genetic effect   A health effect that shows up in subsequent generations.

glass-fiber filter   Filters made of uniform glass fibers with pore sizes 0.7 to 2.7 μm. Used to filter fine particles and algae while maintaining a high flow rate.

Gooch crucible   See *filtering crucible.*

grab sample   A single water sample collected at one time from a single point.

graduated cylinder   A tall, cylindrical glass or plastic container with quantity graduation marks on the side and a pouring lip, used for measuring liquids quickly without great accuracy.

gravimetric procedure   Any analytical procedure that uses the weight of a constituent to determine its concentration.

groundwater under the direct influence of surface water (GWUDI)   A term used in state and federal regulations to designate groundwater sources that are considered vulnerable to contamination from surface water. Systems using such sources must generally provide monitoring and treatment as if they were using a surface water source.

GWUDI   See *groundwater under the direct influence of surface water.*

HAA5   Total concentration of the five haloacetic acids. See also *halocatic acids.*

half-life (radioactive)   The time required for one half of a radioactive isotope to decay.

haloacetic acids   Chemicals formed as a reaction of disinfectants with contaminants in water, consisting of monochloroacetic acid, dichloroacetic acid, trichloroacetic acid, monobromoacetic acid, and dibromoacetic acid.

hardness   A characteristic of water caused primarily by the salts of calcium and magnesium. Causes deposition of scale boilers, damage in some industrial processes, and sometimes objectional taste; may also decrease the effectiveness of soap.

herbicide   A compound, usually a synthetic organic chemical, used to stop or retard plant growth.

heterotrophic plate count (HPC)   A laboratory procedure for estimating the total bacterial count in a water sample. Also called *standard plate count, total plate count,* or *total bacterial count.*

hot plate   An electrical heating unit used in a laboratory to heat solutions.

HPC   See *heterotrophic plate count.*

ICR   See *Information Collection Rule.*

incubate   To maintain microorganisms at a temperature and in an environment favorable to their growth.

incubator   A heated container that maintains a constant temperature for development of microbiological cultures.

indicator   A chemical solution used to produce a visible change, usually in color, at a desired point in a chemical reaction, generally a prescribed end point.

indicator organisms   Microorganisms whose presence indicates the presence of fecal contamination in water.

Information Collection Rule (ICR)   A federal regulation requiring large water systems to collect special information to build up a database that will assist in the development of new monitoring and treatment regulations.

inorganic chemical   A chemical substance of mineral origin not having carbon in its molecular structure.

insecticide   A compound, usually a synthetic organic chemical, used to kill insects.

IOC   See *inorganic chemical.*

iodometric method   A procedure for determining the concentration of dissolved oxygen in water, also known as the *modified Winkler method.*

iron bacteria   Bacteria that use dissolved iron as an energy source. They can create serious problems in a water system because they form large masses that clog well screens, pumps, and other equipment.

ion exchange resin   Beadlike material that removes ions from water, used in deionizers.

isotopes   Varieties of the same element with different masses (different number of neutrons).

jar test apparatus   An automatic stirring machine equipped with three to six paddles and a variable-speed motor drive. Used to conduct the jar test for evaluating the coagulation, flocculation, and sedimentation processes.

Langelier saturation index (LSI)   A numerical index that indicates whether calcium carbonate will be deposited or dissolved in a distribution system. The index is a general indicator of the corrosivity of water.

magnetic stirrer   A device used for mixing chemical solutions in the laboratory.

maximum contaminant level (MCL)   The maximum permissible level of a contaminant in water as specified in the regulations of the Safe Drinking Water Act.

maximum contaminant level goal (MCLG)   Nonenforceable health-based goals published along with the promulgation of an MCL. Originally called *recommended maximum contaminant levels (RMCLs).*

maximum residual disinfectant level (MRDL)   The maximum free chlorine, chloramine, and chlorine dioxide residual allowable in distribution system water.

MCL   See *maximum contaminant level.*

MCLG   See *maximum contaminant level goal.*

membrane filter   A filter made of cellulose acetate with a uniform, small pore size. Used for microbiological examination.

membrane filter (MF) method   A laboratory method used for coliform testing. The procedure uses an ultrathin filter with a uniform pore size smaller than bacteria—less than a micron. After water is forced through the filter, the filter is incubated in a special media that promotes the growth of coliform bacteria. Bacterial colonies with a green-gold sheen indicate the presence of coliform bacteria.

meter   An instrument (usually electronic) used to measure water quality parameters such as pH.

methyl orange   An indicator used in the measurement of the total alkalinity of a water sample.

mg/L   See *milligrams per liter.*

microbiological   Relating to microorganisms and their life processes.

milk dilution bottle   See *dilution bottle.*

milligrams per liter (mg/L)   A unit of the concentration of a water or wastewater constituent: 0.001 g of the constituent in 1,000 mL of water. In reporting the results of water and wastewater analysis it has generally replaced parts per million, to which is approximately equivalent.

MMO–MUG technique   An approved bacteriological procedure for detecting the presence or absence of total coliforms.

modified Winkler method   A modification of the standard Winkler (iodometric) method that uses an alkali-iodide-azide reagent to make the procedure less subject to interferences.

Mohr pipette   A pipette with a graduated stem used to measure and transfer liquids when great accuracy is not required.

monitoring   Routine observation, sampling, and testing of water samples taken from different locations within a water system to determine water quality, efficiency of treatment processes, and compliance with regulations.

mottled   Spotted or blotched. Teeth can become mottled if excessive amounts of fluoride are consumed during the years of tooth formation.

MRDL   See *maximum residual disinfectant level.*

muffle furnace   A high-temperature oven used to ignite and burn volatile solids, usually operated at temperatures near 600°C.

multiple-tube fermentation (MTF) method   A laboratory method used for coliform testing, which uses a nutrient broth placed in culture tubes. Gas production indicates the presence of coliform bacteria.

mutagen   A substance that can change the structure of deoxyribonucleic acid and thus change the basic blueprint for cell replication.

National Primary Drinking Water Regulations (NPDWRs)   Regulations developed under the Safe Drinking Water Act. The regulations establish maximum contaminant levels, monitoring requirements, and reporting procedures for contaminants in drinking water that endanger human health.

natural radioactive series   A sequence of elements that exist naturally and decay into each other in a serial fashion.

negative sample   When referring to the multiple-tube fermentation or membrane filter test, any sample that does not contain coliform bacteria. Also called *absence.*

nephelometer   An instrument that determines turbidity be measuring the amount of light scattered by turbidity in a water sample. It is the only instrument approved by the US Environmental Protection Agency to measure turbidity in treated drinking water.

nephelometric turbidimeter   See *nephelometer.*

nephelometric turbidity unit (ntu)   The amount of turbidity in a water sample as measured using a nephelometer.

neurotoxic   Having a poisonous effect on nerve tissue.

nomographic method   A method that uses a graph or other diagram to solve formulas and equations.

nontransient, noncommunity public water system   A system having its own water supply and serving an average of at least 25 persons who do not live at the location but who use the water for more than 6 months per year.

NPDWRs   See *National Primary Drinking Water Regulations.*

ntu   See *nephelometric turbidity unit.*

Office of Ground Water and Drinking Water (OGWDW)   The office within US Environmental Protection Agency having responsibility for the administration of the Safe Drinking Water Act.

OGWDW   See *Office of Ground Water and Drinking Water.*

opportunistic bacteria   Several types of bacteria that are not usually a danger to persons in good health but can cause sickness or death in persons who are in a weakened condition.

organic chemical   A chemical substance of animal or vegetable origin having carbon in its molecular structure.

oven   A chamber used to dry, burn, or sterilize materials.

oxidant   Any chemical substance that promotes oxidation.

oxidize   To chemically combine with oxygen.

P–A test   See *presence–absence test.*

parts per million (ppm)   The number of weight or volume units of a constituent present with each one million units of the solution of mixture. Formerly used to express the results of most water and wastewater analyses but being replaced by milligrams per liter. For drinking water analysis, concentrations in parts per million and milligrams per liter are equivalent.

pathogens (pathogenic)   Disease-causing organisms.

pCi   See *picocurie.*

petrie dish   A shallow glass or plastic dish with vertical sides, a flat bottom, and a loose-fitting cover. Used for growing microbiological cultures.

pHs   A measure of water's acidity or alkalinity. A scale of 0 to 14 is used, with 0 being extremely acidic and 14 being extremely alkaline.

phenanthroline method A colorimetric procedure used to determine the concentration of iron in water.

phenolphthalein indicator A chemical color-changing indicator used in several tests, including tests for alkalinity, carbon dioxide, and pH.

pHs The theoretical pH at which calcium carbonate will neither dissolve nor precipitate. Used to calculate the Langelier saturation index.

pH meter A sensitive voltmeter used to measure the pH of liquid samples.

photometer An instrument used to measure the intensity of light transmitted through a sample or the degree of light absorbed by a sample.

picocurie (pCi) The measurement of radioactivity most often used in drinking water standards, equal to 10–12 Ci.

pipette Slender glass or plastic tube used to measure and transfer small volumes (usually less than 25 mL) of liquids.

platinum–cobalt method A procedure used to determine the amount of color in water.

PN See *public notification.*

point-of-use (POU) treatment A water treatment device used by a water customer to treat water at only one point, such as at a kitchen sink. The term is also sometimes used interchangeably with "point-of-entry treatment" to cover all treatment installed on customer services.

positive sample In reference to the multiple-tube fermentation or membrane filter test, any sample that contained coliform bacteria. Also called *presence.*

potentiometric method Any laboratory procedure that measures a difference in electric potential (voltage) to indicate the concentration of a constituent in water.

POU treatment See *point-of-use treatment.*

ppm See *parts per million.*

precipitate To separate a substance from a solution or suspension by a chemical reaction.

precursor compound Any of the organic substances that react with chlorine and other disinfectors to form trihalomethanes and other disinfection by-products.

presence   See *positive sample.*

presence–absence (P–A) test   An approved bacteriological procedure for the detection of total coliforms. The results are qualitative rather than quantitative.

presumptive test   The first major step in the multiple-tube fermentation test. The step presumes (indicates) the presence of coliform bacteria on the basis of gas production in nutrient broth after incubation.

primary enforcement responsibility (primacy)   The acceptance by states of the responsibility for enforcing the Safe Drinking Water Act requirements.

probe method   See *electrode method.*

progeny   The various new elements that are formed as a result of transmutation of a radioactive substance.

protozoa   Small, single-cell animals, including amoebas, ciliates, and flagellates.

public notification (PN)   A required notice to the public given by water systems that violate operating, monitoring, or reporting requirements.

public water system (PWS)   As defined by the Safe Drinking Water Act, any system, publicly or privately owned, that serves at least 15 service connections, 60 days out of the year, or serves an average of 25 people at least 60 days out of the year.

PWS   See *public water system.*

QA   See *quality assurance.*

QC   See *quality control.*

quality assurance (QA)   A program to ensure consistency in analytical results between laboratories by periodically testing each laboratory through the analyses of a precisely prepared blind sample.

quality control (QC)   A laboratory program of continually checking techniques and calibrating instruments to ensure consistency in analytical results.

rad   A measure of the dose absorbed by the body from radiation (100 ergs of energy in 1 g of tissue). The abbreviation stands for radiation absorbed dose.

radioactive decay   A process by which the nucleus of an atom transforms to a lower energy state by emitting alpha, beta, or gamma radiations.

radionuclide   A material with an unstable atomic nucleus that spontaneously decays or disintegrates, producing radiation.

reagent bottle   A bottle made of borosilicate glass fitted with a ground-glass stopper, used to store reagents (standard chemical solutions).

recarbonation   The process of adding carbon dioxide as a final stage in the lime–soda ash softening process to convert carbonate to bicarbonates. This process prevents precipitation of carbonates in the distribution system.

receiver   The water treatment system staff member taking information from a customer regarding a water quality complaint.

reg-neg   See *regulatory negotiation process.*

regulatory negotiation process (reg-neg)   A US Environmental Protection Agency process drawing on the experience of many people in the water works field to "negotiate" the various issues in preparing a new draft regulation for public comment.

rem   A quantification of radiation in terms of its dose effect on the human body; the number of rads times a quality factor. The abbreviation stands for *radiation equivalent man.*

representative sample   A sample containing all the constituents that are in the water from which it was taken.

routine (required) sample   A sample required by the National Primary Drinking Water Regulations to be taken at regular intervals to determine compliance with the maximum contaminant levels.

Safe Drinking Water Act   A federal law enacted Dec. 16, 1974, setting up a cooperative program among local, state, and federal agencies to ensure safe drinking water for consumers.

sample bottle   A wide-mouth glass or plastic bottle used for taking microbiological and chemical water samples.

SDWA   See *Safe Drinking Water Act.*

Secondary Drinking Water Regulations   Regulations developed under the Safe Drinking Water Act that establish maximum levels for substances affecting the taste, odor, or color (aesthetic characteristics) of drinking water.

selective absorption   A method used in gas chromatography to separate organic compounds so their concentrations can be determined.

sequestering   A chemical reaction in which certain chemicals (sequestering or chelating agents) "tie up" Other chemicals, particularly metal ions, so that the chemicals no longer react. Sequestering agents are used to prevent the formation of precipitates or other compounds.

single-pan balance   A balance used to make quick, accurate weight measurements. The material to be weighed is placed on the pan, and counterweights, located on arms (beams) beneath the pan, are adjusted to balance the material, thus indicating the weight. Also known as a *beam balance*.

SOCs   See *synthetic organic chemicals*.

SPADNS method   A colorimetric procedure used to determine the concentration of fluoride ion in water. SPADNS (sodium 2-(parasulfophenylazo) 1.8-dihydroxy-3,6-naphthalene disulfonate) is the chemical reagent used in the test.

specific ion meter   A sensitive voltmeter used to measure the concentration of specific ions, such as fluoride in the water. Electrodes designed specifically for each ion must be used.

spectrophotometer   A photometer that uses a diffraction grating or a prism to control the light wavelengths used for specific analysis.

splash goggles   Safety goggles with shatterproof lenses designed to provide a tight covering around the eyes, protecting them from chemicals and flying particles.

stable   Resistant to change.

Surface Water Treatment Rule (SWTR)   A federal regulation established by US Environmental Protection Agency under the Safe Drinking Water Act that imposes specific monitoring and treatment requirements on all public drinking water systems that draw water from a surface water source.

SWTR   See *Surface Water Treatment Rule*.

synthetic organic chemicals (SOCs)   Generally applied to manufactured chemicals that are not as volatile as volatile organic chemicals. Included are herbicides, pesticides, and chemicals that are widely used in industries, such a ethylbenzene, styrene, and toluene.

TCR   See *Total Coliform Rule*.

TD   A mark on a pipette meaning "to deliver." The pipette is calibrated to deliver the calibrated volume of the pipette with a small drop left in the tip.

teratogenic effect   A health effect on a fetus.

test tube   A slender glass or plastic tube with an open top and rounded bottom. Used for a variety of tests.

THMs   See *trihalomethanes.*

threshold odor number (TON)   A number indicating the greatest dilution of a water sample (using odor-free water) that still yields a noticeable odor.

time composite   A composite sample consisting of several equal-volume samples taken at specified times.

titration   A method of analyzing the composition of a solution by adding known amounts of a standardized solution until a given reaction or end point (color change, precipitation, or conductivity change) is produced.

titrimetric method   Any laboratory procedure that used titration to determine the concentration of a constituent in water.

TOC   See *total organic carbon.*

TON   See *threshold odor number.*

Total Coliform Rule (TCR)   A regulation that became effective Dec. 31, 1990, doing away with the previous maximum contaminant level relating to the density of organisms and relating only to the presence or absence of the organisms in water.

total coliform test   Either the multiple-tube fermentation or the membrane filter test. Both tests indicate the presence of the entire coliform group or total coliforms.

total organic carbon (TOC)   The results of a general analysis performed on a water sample to determine the total organic content of the water.

total trihalomethanes (TTHMs)   The total of the concentration of all the trihalomethane compounds found in the analysis of a water sample.

toxic   Causing an adverse effect on various body parts (such as the liver or kidneys).

transect   An imaginary line along which samples are taken at specified intervals. Transect sampling is usually done on large bodies of water such as rivers and lakes.

transfer pipette   See *volumetric pipette.*

transient, noncommunity public water system An establishment having its own water system, where an average of at least 25 persons per day visit and use the water occasionally or for only short periods of times.

transmutation The changes that take place in a radioactive substance due to radioactive disintegration.

trihalomethanes (THMs) A group of compounds formed when natural organic compounds from decaying vegetation and soil (such as humic and fulvic acids) react with chlorine.

TTHMs See *total trihalomethanes.*

turbidimeter An instrument that measures the amount of light impeded or scattered by suspended particles in a water sample, using a standard suspension as a reference.

turbidity A physical characteristic of water that makes the water appear cloudy. The condition is caused by the presence of suspended matter.

unreasonable risk to health (URTH) A determination that must be made before a variance or an exemption can be granted. Factors that must be considered in making the determination that there is no unreasonable risk to health include the degree to which the MCL is exceeded, the adverse health effects involved, the duration of the exceedance that is expected, how long the problem has existed, and the type of population exposed.

URTH See *unreasonable risk to health.*

US Environmental Protection Agency (USEPA) A US government agency responsible for implementing federal laws designed to protect the environment. Congress has delegated implementation of the Safe Drinking Water Act of USEPA.

USEPA See *US Environmental Protection Agency.*

USPHS See US Public Health Service.

US Public Health Service (USPHS) A government agency that established early standards for acceptable drinking water quality under provisions of the Interstate Quarantine Act of 1893.

utility oven A laboratory oven used primarily to dry labware and chemicals prior to weighing or to sterilize labware.

vacuum pump A pump used to provide a partial vacuum, needed for filtering operations such as the membrane filter test.

viable   Capable of living.

VOCs   See *volatile organic chemicals.*

volatile organic chemicals (VOCs)   Lightweight organic compounds that vaporize easily.

volumetric flask   A squat bottle with a long, narrow neck, used to prepare fixed volumes of solution. Each flask is calibrated for a single volume only.

volumetric pipette   A pipette calibrated to deliver a single volume only.

water still   A device used to produce distilled water by evaporation and condensation of tap water.

waterborne disease   Any illness caused by a pathogenic organism carried by water.

zeta potential   The resistance between suspended particles in water.

Index

NOTE: *f.* indicates a figure; *t.* indicates a table.

Printed in the United States
85121LV00003B/131-500/A